THE ROUGH GUIDE to the

iPhone

Peter Buckley

www.roughguides.com

Credits

The Rough Guide to the iPhone

Text, design and layout:
Peter Buckley & Duncan Clark
Proofreading: Susanne Hillen
Production: Rebecca Short

Rough Guides Reference

Editors: Kate Berens,
Tom Cabot, Tracy Hopkins,
Matt Milton, Joe Staines
Director: Andrew Lockett

Apple hardware images courtesy of Apple UK

Acknowledgements

The author would like to thank everyone at Rough Guides, above all Sean Mahoney, who gallantly stood in line for 14 hours on the day of the iPhone's launch to secure the hardware used to write the first edition of this book. Also Andrew Lockett and John Duhigg for signing up the project in record time.

Publishing information

This third edition published September 2010 by
Rough Guides Ltd, 80 Strand, London WC2R 0RL
11, Community Centre, Panchsheel Park, New Delhi 110017, India
mail@roughguides.com

Distributed by the Penguin Group:
Penguin Books Ltd, 80 Strand, London WC2R 0RL
Penguin Group (USA), 375 Hudson Street, NY 10014, USA
Penguin Group (Australia), 250 Camberwell Road, Camberwell, Victoria 3124, Australia
Penguin Group (New Zealand), 67 Apollo Drive, Mairangi Bay, Auckland 1310, New Zealand

This paperback edition published in Canada in 2010. Rough Guides is represented in Canada by Tourmaline Editions Inc., 662 King Street West, Suite 304, Toronto, Ontario, M5V 1M7

Typeset in Minion and Myriad
Printed and bound in the USA by Lake Book Manufacturing

272 pages; includes index

A catalogue record for this book is available from the British Library.

ISBN: 978-1-84836-641-1

1 3 5 7 9 8 6 4 2

Contents

00 Introduction 5

Primer

01 FAQs 11
02 Buying an iPhone 31

Basics

03 Getting started 39
04 Typing tips 58
05 Connecting 64
06 Apps & webclips 71
07 Maintenance 83

Phone

08 Contacts 95
09 Calls 109
10 SMS 130

iPod

11 iTunes prep 137
12 The iTunes Store 149
13 iPhone audio 156
14 iPhone video 167

Internet

15 The web 175
16 Email 187

More...

17 Camera & photos 201
18 Maps 209
19 Reading 215
20 More tools 224

iPhonology

21 Accessories 235
22 Websites 246
23 iPhone weirdness 249

Index 261

00

Introduction

The iPhone was probably the most keenly anticipated consumer product of all time. After years of rumours, predictions and discussions, Apple finally previewed the device in early 2007. That was six months before the iPhone went on sale in the US and nearly a year before it would be available in Europe, but photos of the gadget were nonetheless all over newspaper front pages around the world the following morning. It was the sort of media coverage that no other company could even have dreamed of. But then no other company had created the iPod.

In the months after the preview, traders sold iPhone-related web addresses on eBay for huge sums, accessory manufacturers scrambled to produce add-ons for a device they'd never actually seen, and bloggers dissected Steve Job's presentation in minute detail for clues about what the iPhone would or wouldn't be able to do.

By the time the phone actually went on sale, things had reached fever pitch. Apple and AT&T stores across the US saw lines of early adopters camping on the streets to ensure that they wouldn't miss out on release day and have to wait a couple of weeks for restocks. Savvy students hired themselves out as line-standers, and websites pointed iPhone campers in the direction of the most conveniently located toilets, restaurants and DVD rental stores.

That was the Friday. By the end of the weekend, more than half a million phones had been sold, with another half-million or so flying off the shelves the following week.

But was all the excitement justified? Not according to some commentators, who described such scenes as examples of mass hysteria – triumphs of PR spin. Phones that do email, music and web browsing had existed for years, they pointed out, and many of them offered greater functionality or a broader feature array than the iPhone. And at a lower price, too.

But that was to miss the point. There had indeed been many smartphones before the iPhone, but they were consistently ugly, fiddly, counter-intuitive and unergonomic. They looked and felt as if they were designed by people interested in dry functionality rather than whether a device is actually pleasurable to use – people who create "solutions" rather than everyday products. Such phones were aimed primarily at businesses and had names reminiscent of the corporate computer systems they were designed to work with – the Motorola MC70, for instance, or Palm Treo 700WX.

By comparison, the iPhone was, from day one, nicely designed, simple to use, and – for the most part – did what you actually wanted it to do. Its innovations are not in its feature array but in its human interface – a screen that lets you easily flick through long lists of contacts or zoom in on a webpage with the touch of a finger or two; a sensor that lets you rotate the device to make a landscape picture fill the screen or control a video game.

Once people had seen these features in action, the hype was generated from the bottom up. All Apple did was to put out a few ads – there was no guerilla marketing campaign or PR offensive. (Indeed, as all technology journalists know, it's difficult to even get through to Apple PR, let alone get someone to return your call or tempt your goodwill with freebies.)

As if the prove the point, the bottom-up buzz never went away. In June 2010 the launch of the fourth-generation iPhone saw punters queuing outside stores once again, with over three million units being sold in the first few weeks alone. Even the accompanying "Antenna-gate" media circus (surrounding alleged reception issues when the phone is held in a certain way) failed to crush the enthusiasm of the majority of iPhone owners.

None of which is to say that the iPhone is beyond criticism. It has its faults and limitations, of course. And you could make a reasonable case that so much buzz over a mobile phone, however pretty and ergonomic it may be, is an example of rampant, unsustainable consumerism – a focus on expensive toys in an era of conflict and climate change.

That may well be, but mobile phones are here to stay. And if we're going to have them, we may as well have ones that are good to use, long-lasting, include iPod functions, and can take great pictures, so that we don't end up buying and carrying around multiple devices. The iPhone checks all of these boxes, and plenty more besides.

About the book

Whether you're thinking about buying an iPhone or already have one, this book is for you. It covers the whole topic, from questions you might want answering before you buy, and advice on importing numbers from an old handset, through to advanced tips and tricks and pointers towards some of the apps that are definitely worth downloading from the iTunes App Store.

We've even included, towards the back of the book, some samples of the spoofs and other weird and wonderful online creations that the iPhone phenomenon has inspired. However much of an iPhone addict you become, it's good to know that some people are even more obsessed...

This book was written using a fourth-generation iPhone running iPhone OS 4.0.1. (To see which version of the software you're running, connect your iPhone to iTunes, highlight it in the iTunes sidebar, and then click the Summary tab.) If you're running a later version of the OS then you may well come across features not covered here, though the majority of what's written will still apply. If you have an earlier version of the software or an older iPhone, you may be missing a few of the features described here but, again, the majority of the book will still be relevant.

Primer

01

FAQs

Everything you ever wanted to know but were afraid to ask

The big picture

What's an iPhone?

An iPhone is a smartphone – in other words, a mobile phone that doubles as a handheld computer, complete with web browsing, email, music playback and the ability to run applications, or "apps". The device is produced by Apple, manufacturers of the iPod, iPad and Mac computers.

After years of speculation, and to great media fanfare, the iPhone was launched in the US and UK in 2007, with an updated version released each summer since.

How does it compare to other smartphones?

There are lots of smartphones on the market, many of which offer a similar feature list to that of the iPhone, including web browsing, email and downloadable apps. One key difference is the

user-interface: the iPhone is (generally) simple and pleasurable to use, whereas many other smartphones are fiddly and confusing. However, the iPhone also offers various other benefits, relative to its competitors:

- A huge, bright, super-high-resolution screen that changes orientation when you rotate the phone.
- A flexible touch-screen interface.
- An excellent web browser.
- Compatibility with iTunes.
- Wi-Fi and 3G capability for fast Internet access.
- Visual Voicemail, which lets you hear messages in any order.
- Loads of downloadable games and other apps.
- High-quality video calls.

Of all these benefits, perhaps most significant is the sheer number of third-party apps available for the iPhone. These allow you to add almost any function that Apple missed off.

What about Android phones?

Android is the Google software that runs on many non-Apple smartphones, made by a diverse number of hardware manufacturers, each of which tweak the software for their own needs; as such, some Android phones tend to be "better" than others. Whether you choose an iPhone or Android phone is really down to personal preference. Interestingly, the rivalry between the two platforms and their users is beginning to mimic the rivalry between Mac and Windows PC users in the home computing space. For the full story, check out *The Rough Guide to Android.*

What's an app?

App is short for application – a piece of software designed to fulfil a particular function. If you're used to a PC, an app is basically the same as a program. On the iPhone an app might be anything from a game, a word processor or a retro-styled alarm clock through to a version of a popular website with extra features added specially for the iPhone.

At the time of writing, there are hundreds of thousands of apps available to download from the App Store (see p.72). Some of these apps have been downloaded and installed on millions of iPhones.

How does the iPhone compare to the iPod?

The iPhone *is* an iPod, in addition to being a phone and an Internet device. It can do nearly everything an iPod can do: play music, videos and podcasts, display album artwork using "Cover Flow" (see p.159), and create playlists based on a single track using the "Genius" feature (see p.161).

The main advantage of a traditional iPod is storage space. At the time of writing, iPhones offer 16 or 32 gigabytes of space, while the iPods offer up to 160 gigabytes.

How does it compare to a computer for web browsing and email?

No pocket-sized Internet device can match a computer with large screen, mouse and a full-sized keyboard, but the iPhone comes about as close as you could hope. Web browsing, in particular, is very well handled, with a great interface for zooming in and out on sections of a webpage. However, when you're out and about, you'll find the access slow compared to a home broadband connection. (More on this later.)

Can it open and edit Word and Excel docs?

Out of the box, the iPhone can open, read and forward Word, Excel and PowerPoint docs sent by email or found online, but it doesn't allow you to actually edit them. To do this, you'll need to download a suitable app.

What's iOS and OS X?

All computers have a so-called operating system – the underlying software that acts as a bridge between the hardware, the user and the applications. The standard operating system on PCs is Windows; on Macs, it's OS X.

The operating system on the iPhone – known as iOS – is based on OS X, though you wouldn't know it, because it's slimmed down and looks and feels specially designed to make the most of the phone's small touch-screen interface.

Is the on-screen keyboard easy to use?

Apple are very proud of the iPhone's touch-screen keyboard and the accompanying error-correcting software that aims to minimize typos. In general, reviewers and owners alike have been pleasantly surprised at how quickly they've got used to it.

Inevitably, however, it's not to everyone's taste. When using many applications you can rotate the iPhone through ninety degrees to use a bigger version of the keyboard in landscape mode.

Do I need a computer to use an iPhone?

Yes – or at the very least you need access to one. You can't activate an iPhone without a Mac or PC. Moreover, a computer is one way to get music, video and other media onto the phone. To copy a CD onto an iPhone, for example, you first copy it onto the computer and then to the phone. However, it's also possible to download music and video files straight to the iPhone from the iTunes Store (see p.72).

Is my current computer up to the job?

If you bought a PC or Mac in the last few years, it will probably be capable of working with an iPhone, but it's certainly worth checking before spending any money. The box overleaf explains the minimum requirements and the upgrade options if your computer doesn't have what it takes.

How does the phone connect to the computer?

With a USB2 cable either connected directly to the iPhone or to a Dock (see p.46) into which the iPhone slots. The phone can't currently sync with iTunes wirelessly.

What's iTunes?

iTunes is a piece of software produced by Apple for Macs and PCs. It serves four main functions:

- **Media manager** iTunes is used for managing and playing music, video and other media on a Mac or PC. This includes everything from importing CDs to making playlists.

• **iPhone, iPod or iPad manager** iTunes is the program that is used for copying music, movies, photos and other data from a computer to Apple mobile devices, including the iPhone, iPod and iPad.

• **Download store** The iTunes Store is Apple's legal, pay-to-use music, video and TV-show download service. It's built into

Computer requirements & upgrade options

As of June 2010, to use the most recent iPhone model (v4) you need a computer that meets the following minimum requirements:

Mac: OS X 10.5 or later and a USB2 port
PC: Windows 7, Vista or XP and a USB2 port

Here's how to check if your computer fulfils these criteria:

Checking your version of Windows

To check your operating system on a PC, right-click the My Computer icon and select Properties. If you don't have Windows Vista or XP, you probably have Windows Me or 98. You could consider upgrading, but you'll have to pay (around $95/£85); it can be a bit of a headache (some older programs and hardware won't work on XP or Vista); and you'll need to check whether your hardware is up to the job (see microsoft.com for more details). If you have Windows 95, it's probably a matter of buying a new computer.

Also note that, if you have an older copy of Windows XP, you may need to run Windows Update to install the Service Pack 3 update before your computer will work with an iPhone.

Checking your version of Mac OS X

On a Mac, select About This Mac or System Profiler from the Apple menu in the top-left corner of the screen. If the OS X number starts with 10.4 or 10.5, you'll be fine, though you might have to run the Software Update tool (also in the Apple menu) to grab the latest updates.

If you have OS X 10.4 or earlier, the latest iPhone won't work; you'll have to upgrade to the latest version of OS X, which will cost you around $130/£90. But first check the hardware requirements to make sure that your machine can handle the newer operating system (see apple.com/macosx) and also iTunes as well as into the iPhone itself and has sold billions of tracks. It also offers access to the App Store, allowing you to browse and buy applications from your Mac and PC.

• **Podcast aggregator** You can use iTunes to subscribe for free to hundreds of audio or video podcasts for use on your computer, iPhone or iPod.

check whether you have a USB2 socket (see below). You may find that you'd be better off buying a new Mac.

USB2 port

This is the socket to which you connect the iPhone. Any USB socket will work, but data transfer to the phone will be painfully slow unless you have a USB2 socket, which is why Apple specify USB2 as a minimum requirement. USB2 sockets are found on most PCs and Macs made after 2003; to be sure, check your computer's manual, as USB and USB2 ports look identical.

If you don't have a USB2 port, you could add one to your computer by purchasing a suitable adapter. These are available for around $30/£20 for desktops and around double that for laptops. Or, if you have the right port but it's usually occupied by another device, buy a powered "hub" to turn a single port into two or more. These start at around $30/£20.

iTunes

On both a Mac and PC you will also want to make sure you have the most up-to-date version of the iTunes software before you get started (see p.40).

I heard there was controversy about the battery. What's that about?

Like the original iPods, the iPhone has received quite a lot of negative publicity over its battery, which gradually dies over time and can only be replaced by Apple for a substantial fee ($79/£63 at the time of writing).

To be fair, however, *all* rechargeable batteries deteriorate over time and eventually die. The only difference with iPhones is that the replacements cost more than some other phones – and that you can't fit them yourself (at least not in theory). The high cost can partly be explained by the nature of the battery. Few pocket devices offer hours of video playback or such large, bright screens. That type of performance is comparable to a laptop – and laptop batteries are even more expensive. As for sending the device back to Apple, this is irritating, for sure, though at least it means that the device can be properly sealed and be free from the flimsy battery flaps that often get broken on other phones.

Though Apple don't recommend it, it's possible to source bargain iPhone batteries on the Internet and fit them yourself, or pay someone else to do it. As for the longevity of the battery, expect to replace it after 2 to 4 years, depending on how much you drain it each day. For more information about the battery, see p.87.

Does the iPhone contain a hard drive?

No. Like the iPod Nano, Touch and Shuffle, the iPhone stores its information on so-called flash memory: tiny chips of the kind found in digital camera memory cards and flash drives. These have a smaller capacity than hard drives, but on the other hand they're less bulky, less power hungry and less likely to break if the device is dropped.

Does the screen scratch easily?

It's not invulnerable, but the iPhone screen is significantly more scratch resistant than that of an iPod – especially in the case of the most recent model, the iPhone 4. Nonetheless, you might want to consider investing in an inexpensive screen-protecting film (see p.238) or some kind of wrap-around case (see p.239) to protect the screen when your phone is in a pocket or bag.

Does the iPhone present any security risks?

Not really. There's a theoretical risk with any smartphone or computer that someone could "hack" it remotely and access any stored information. But the risk is extremely small – especially in the case of the iPhone which, by default, won't allow Bluetooth access from laptops or other phones. The only real risk – as with any phone – is that someone could steal your phone and access your private data or make expensive long-distance calls on your account. If you're worried about that possibility, the best defence is to password-protect your iPhone – see p.55 to find out how.

Is it possible to type with one hand?

Yes, this is perfectly possible: your fingers hold the body of the device and your thumb can access the screen. However, typing is faster with two hands.

Will the touch screen work with gloves on?

No – unless, of course, you wear fingerless gloves. But then most modern mobiles are at best fiddly with a pair of gloves. You can, however, use certain brands of stylus with an iPhone; the pick of the bunch is the Pogo Stylus (tenonedesign.com/stylus). You could also take your lead from Korean iPhone users, who swear by a particular brand of snack sausage (tinyurl.com/yldwb38).

Phone issues

If I get an iPhone, do I have to change network and get a new contract?

It depends where you are. At the time of writing, the iPhone is available exclusively on AT&T in the US, whereas in the UK it is sold unlocked and can be used on any participating network, including O2, Orange and Vodafone. If you're already with a participating network, you can simply add an unlimited data plan (for web browsing and email) to your existing voice plan. Expect to pay around $20/£10 extra per month.

If your existing network doesn't offer iPhone plans, you'll need to switch. If you're tied in to your existing contract for a set period of time, this might mean paying a get-out fee.

Locked in with another carrier?

If you're locked into a contract with a phone company that doesn't offer the iPhone, there are various things you could try in order to excuse yourself from the get-out fee. Soon after the iPhone's launch, Wired.com's Daniel Dumas listed a whole host of ideas, from faking your own death to being deployed abroad for military service. Here are a few of his more realistic suggestions:

- **Hand on your contract** Use a website such as CellTradeUSA.com to find someone willing to take your plan off your hands.

- **Move out of range** Tell your phone company that you're moving to an area that their network doesn't cover and they may terminate the contract.

- **Beat them at their own game** Check your original contract to see whether it would be rendered void by small changes they've already made, such as giving you extra minutes or text messages.

- **Ask to see the contract** The carrier should be able to show you the contract that you signed. If they can't – or they simply can't be bothered to dig it out from their records office – that may give you grounds to walk.

Either way, you'll get to keep your current phone number; transferring the number is part of the activation process.

Can I use my current SIM card?

If moving to an iPhone 4, the answer is almost certainly no, as this model uses a special, smaller format MicroSIM. You may also discover, with any iPhone model, that your current SIM does not comply with the iPhone contracts on offer from a specific network, in which instance any network offering iPhone tariffs will be able to provide you with a new SIM and transfer your number from your old SIM when you sign up. If you insert a non-approved SIM from a different phone, you'll get an error message and it won't work.

Naturally, it didn't take geeks long to figure out how to "unlock" the iPhone for use with any SIM card and network. Sites such as iphoneunlocking.org.uk offer downloadable software to unlock an existing iPhone, either before or after it's been activated. It's also possible to buy unlocked iPhones from non-approved networks in many countries.

There are downsides to unlocking, however. Most obviously, you'll invalidate your warranty. Certain unlocking solutions may also disable some functions and make it impossible (or risky) to install future iPhone software updates from Apple.

What about "jailbreaking"?

Jailbreaking is the process of adapting the iPhone's operating system to allow it to run third-party software that's not approved by Apple. Due to the existence of the App Store (see p.72), and the wealth of applications that it offers, there isn't a great deal of reason to jailbreak an iPhone, but if you're really keen, there's plenty of information available online.

Do I really need to sign a long contract? What about prepaid/pay-as-you-go?

Depending on where you live and your network, you'll probably find you have to enter into a 12-, 18- or 24-month contract as part of the process of getting your iPhone. In the UK, prepaid plans are available. In the US, however, the only way to get a prepaid account is to have a bad credit rating. However, some bloggers with good credit ratings claim to have successfully signed up by entering a social security number of 999-99-9999 during the activation process.

Can I use the iPhone overseas?

Yes, the iPhone is quad-band and will work for voice calls in nearly 200 countries. Overseas calls work automatically for most UK-based customers, but US-based iPhone owners may need to activate international roaming by calling AT&T on 1-800-331-0500. This service is subject to your credit rating or payment history – which reflects the fact that overseas calling can quickly become expensive (see box opposite).

When it comes to foreign use of web, email, maps and other Internet-based features, you're best sticking to Wi-Fi hotspots where you can get online for free (or, at most, the cost of accessing the hotspot). In many countries it's possible to connect via the local phone network as well, but be prepared for some really savage fees. When you're paying up to $19 (AT&T) or £6 (O2) per megabyte, then a week or so of using email, web or maps can quickly add up to hundreds of dollars or pounds. If you can stomach such costs, turn Data Roaming on within Settings > General > Network.

To avoid roaming charges, you could consider trying to get your iPhone unlocked (see p.21), which would allow you to use it with a local pay-as-you-go SIM card from the country you're visiting.

Sample roaming charges

For US iPhone users

These are the fees levied on US AT&T customers using their phones abroad. For a full list of World Travel costs, visit AT&T.com

	Standard rate	World Traveler rate ($5.99 per month)
Calls		
Mexico	$0.99	$0.59
UK, France, Italy, Germany, Spain	$1.29	$0.99
Jamaica, Barbados, T&T	$1.99	$1.69
Brazil, Argentina, China, Turkey	$2.29	$1.99
South Africa, India, Egypt,	$2.49	$1.69–2.49
Russia, Mali, Ukraine, Vietnam	$3–5	$2–5
Messages		
Text messages (charges only applied for sending, not receiving)	$0.50	$0.50
Data use		
Internet/email	$19.50 per MB	

For UK iPhone users

The following rates are applies by O2, as of mid-2010. For the latest prices, see o2international.co.uk

Zones	Call the UK (per min)	Call in zone (per min)	Call out of zone (per min)	Receive call (per min)	Send a text	Photo/ video text
W. Europe	35p	35p	176p	18p	11p	25p
E. Europe	137p	137p	199p	85p	30p	25p
USA & Canada	137p	137p	199p	103p	25p	25p
Asia & Pacific	85p	85p	199p	111p	30p	25p
Other	170p	199p	199p	141p	40p	25p

But unlocking presents various problems, as already discussed, and when a foreign SIM is in place you won't be able to receive calls to your usual number. All told, if you travel to one country a lot, it might be simpler to buy an inexpensive prepaid phone in that country and arrange to have calls to your regular iPhone number forwarded (see p.123).

Will I be able to get the contact numbers off my old phone and onto the iPhone?

In general, yes, but the process depends on your old phone and your computer. See p.96 for more information.

Internet issues

Does the iPhone offer fast Internet access?

That depends where you are. If you're within range of a Wi-Fi network, as found in homes, offices, cafés, etc (and across some entire city centres), then your Internet access will typically be pretty speedy: not quite as fast as with a Mac or PC, but not a long way off.

If, on the other hand, you're out and about – walking down the street, say, or sitting on a bus – then the iPhone will usually have to access the Internet via a mobile phone network. Where you can get access to a 3G signal, the speed of your connection will be pretty good, though in most areas it's still nowhere near as fast as a home broadband connection. Whenever 3G isn't available, your iPhone will try to connect to the slower EDGE and GPRS networks (see opposite). The first-generation iPhone, can't access 3G at all, and so has to rely on EDGE and GPRS: fine for checking emails, but potentially painful if you're browsing picture-heavy websites.

Believe it or not, another factor worth knowing about when out and about is your own speed of travel. On a train or in a moving car, you might find that your connection speed is slower than when stationary. This is because the device is having to accommodate a constantly shifting relationship to the nearby signal masts that it's connecting with, making it hard for the iPhone to maintain a coherent stream of data to and from the Internet.

EDGE, 3G … what's all that about?

Over time, the technology used to transmit and receive calls and data from mobile phones has improved, allowing greater range and speed. Of the network technologies widely available at present, 3G (third generation) is the most advanced, allowing Internet access at speeds comparable to home broadband connections.

A 3G-capable iPhone (such as the iPhone 3G, 3GS and iPhone 4) should automatically connect to your carrier's data network. When it connects to the Internet using the cellular data network, one of three icons will appear in the Status Bar: 3G (which is fastest), E (for EDGE), or ° (for GPRS, which is even slower than EDGE). Cellular data networks will automatically give way to Wi-Fi (which is usually faster) whenever possible.

Much to the surprise of many commentators, the first-generation iPhone didn't access the Internet via 3G. Instead, it used a network known as EDGE (Enhanced Data rates for GSM Evolution, to give it its rather grand full name). According to Apple, the main reasons for this decision were that at the time EDGE was more widely available in non-urban areas than 3G, and that EGDE-based phones offered better battery life.

Just as Apple warned, 3G access can indeed put strain on the battery, so iPhones offer the option to disable this faster connection type (under Settings > General > Network) in order to preserve valuable battery life.

Will all websites work on an iPhone?

The iPhone features a fully fledged web browser – Safari – which will work with the overwhelming majority of websites. The main catch is that certain special types of web content won't display:

- **Flash animations**, widely used for banner ads (no great loss there), interactive graphics and a few entire websites.

- **Java applications** Not to be confused with Javascript (which works fine on an iPhone), Java is used for certain online programs such as calendar tools or broadband speed tests.

- **Some music and video** When you're connected via Wi-Fi, some music and video clips will work fine. But don't expect to be able to play Realplayer or Windows Media Player clips.

What's Safari?

Safari is the web browser built into the iPhone. It's a streamlined version of the browser shipped on all Macs sold in the last few years. Just before the release of the iPhone, in mid-2007, Apple released a version of Safari for PC users.

Can the iPhone sync my bookmarks?

Yes. It works with Safari on Macs, and Safari or Internet Explorer on PCs. Firefox isn't currently, by default, supported (see p.178).

Will the iPhone work with my email account?

For a personal email account, almost certainly. The iPhone is compatible with all standard email technologies, such as POP3 and IMAP, and ships preset to work with AOL, Yahoo!, MobileMe and Gmail or Google Mail. If you use Outlook on a PC or Mail on a Mac, the iPhone will even sync your account details so you don't have to set anything up.

The only time you're likely to encounter problems is when setting up a work email account. This may or may not be possible, depending on the policies of your network administrators. The only way to be sure is to ask. For more on setting up and using email, see p.187.

Can I make phone calls while on the Internet?

Yes, but only when your iPhone is connected to the Internet via a Wi-Fi network. You can't make calls and get online via EDGE or 3G simultaneously.

Can I use the iPhone to get my laptop online when out and about?

Yes, with a recent iPhone model, this is possible, though in the UK you'll have to pay an additional "tethering" fee to your mobile phone provider to get this to work.

What's MobileMe?

Previously known as .Mac, MobileMe is an Apple subscription service that provides a suite of online tools in return for an annual fee (currently $99/£69). Available to Mac and PC users, the service allows you to keep multiple computers and iPhones in sync with each other. For example, when you add a contact or calendar event on your computer, the new details will be instantly "pushed" to your iPhone over the air waves. On a Mac, this works with iCal, Address Book and Mail; on a PC it works with Outlook.

MobileMe also offers a range of other tools such as one-click photo sharing (from computer or iPhone to a personal web gallery), an easy-to-use website builder, and online file storage for backup or transfer (see p.231). You can try MobileMe free for 60 days. To find out more, visit apple.com/mobileme.

iPod issues

Will the iPhone work with my iPod accessories?

Maybe. The Dock socket on the bottom of the iPhone is the same as the one on an iPod, so iPod accessories that connect via this socket should be able to connect to an iPhone. That doesn't necessarily mean they'll work, however – especially if they are more than a couple of years old.

What's the sound quality like for music?

Pretty well the same as with an iPod. Tracks downloaded from the iTunes Store or imported from disc at the default settings sound marginally worse than CD quality. However, you're unlikely to notice any difference unless you do a side-by-side comparison through high-quality speakers or headphones.

Anyhow, this sound quality isn't fixed. When you import tracks from CD (or record them from vinyl) you can choose from a wide range of options, up to and including full CD quality. The only problem is that better-quality recordings take up more disk space, which means fewer tracks on your phone. The trade-off between quality and quantity is entirely for you to decide upon. For more information, see p.138.

Can I use my existing earphones or headphones?

Yes, with any iPhone model. However, for first-generation iPhones you might need an adapter, since the headphone socket is positioned on the curved corner of the device and recessed in such a way that most regular headphone jacks won't fit.

Aside from this, if you use regular headphones or earphones, you won't be able to use the mic and button that form part of the supplied earbuds. These allow you to answer calls and control music playback without getting the device out of your pocket.

For details of some headphones and adapters that do work with the iPhone, turn to p.236.

I've never had an iPod. Isn't it a hassle to transfer all my music from CD to computer to iPhone?

It certainly takes a while to transfer a large CD collection onto your computer, but not as long as it would take to play the CDs. Depending on your computer, it can take just a few minutes to transfer the contents of a CD onto your computer's hard drive – and you can listen to the music, or work in other applications, while this is happening. Still, if you have more money than time, there are services that will take away your CDs and rip them into a well-organized collection for around $1/£1 per CD.

PodServe podserve.co.uk (UK)
DMP3 dmp3music.com (US)
RipDigital ripdigital.com (US)

Once your music is on your PC or Mac, it only takes a matter of minutes to transfer even a large collection across to the iPhone; and subsequent transfers are even quicker, since only new or changed files are copied over.

Can the iPhone download music directly from the iTunes Store like my computer can?

Yes. For more on the iTunes Store, see p.149.

Is downloading legal?

Yes, as long as you use a legal store such as iTunes. As for importing CDs and DVDs, the law is, surprisingly, still a tiny bit grey in many countries, but in practice no one objects to people importing their own discs for their own use.

What *could* theoretically put you on the wrong side of the law is downloading copyrighted material that you haven't acquired legitimately – and, of course, distributing copyrighted material to other people. A huge amount of music is "shared" illegally using peer-to-peer applications. With millions of people taking part, it seems impossible that everyone will get prosecuted, though there have been a few token subpoenas on both sides of the Atlantic.

What's DRM?

DRM (digital rights management) is the practice of embedding special code in audio, video or eBook files to limit what the user can do with those files. For example, ePub titles downloaded from the iBookstore (see p.215) will only be readable within the iBooks app (on either iPhones, iPads or the iPod Touch). Music and video downloaded from the iTunes Store features what's called FairPlay DRM, which only allows the files to be played on up to five computers that have been authorized to work with the iTunes account used to download them. These files can, however, be used on as many iPhones, iPods and iPads as you want.

02

Buying an iPhone

Which model? Where from?

For most people, buying an iPhone will involve simply walking into a store and picking whichever capacity model they can afford. But, for more cautious buyers, a few questions may need to be answered first. How much storage space do you really need, for example? And would it be better to wait for a next-generation iPhone?

How much space do you need?

At the time of writing, the latest iPhone – version 4 – is available with either 16GB or 32GB of storage space. The amount of space you need depends on the number of songs, photos, movies, podcasts, email attachments and apps you want to be able to store at any one time.

True storage capacity

The first thing you should know is that your iPhone may offer slightly less space than you expect. All computer storage devices are in reality about 7 percent smaller than advertised. The reason is that hardware manufacturers use gigabyte to mean one billion bytes, whereas in computing reality it should be 2^{30}, which equals 1.0737 billion bytes. This is a bit of a scam, but everyone does it and no one wants to break the mould.

Moreover, a few hundred megabytes of the remaining 93 percent of space is used to store the iPhone's operating system, applications and firmware. All told, then, you can expect to lose a decent chunk of space before you load a single video, song or app:

Advertised capacity	Real capacity	Actual free space
32GB	29.8GB	28.4GB
16GB	14.9GB	14.5GB

How big is a gig?

A gigabyte (GB) is, roughly speaking, the same as a thousand megabytes (MB) or a million kilobytes (KB). Here are some examples of what you can fit in each gigabyte.

		1GB =
MUSIC	at 128 kbps (normal quality)	250 typical tracks
	at 256 kbps (high quality)	125 typical tracks
	at 992 kbps (CD quality)	35 typical tracks
AUDIOBOOKS	at 32 kbps	70 hours
PHOTOS		3000 photos
VIDEO		2.5 hours

Checking your current data needs

If you already use iTunes to store music and video, then you can easily get an idea of how much space your existing collection takes up. On the left, click Music, Movies, Podcasts or any playlist and the bottom of the iTunes window will reveal the total disk space each one occupies.

As for photographs, the size of the images on your computer and the amount of space they occupy there bears little relation to the space that the same images would take up on the iPhone. This is because when iTunes copies photos to your phone (see p.206) it resizes them for use on the iPhone's screen. As a guide, 3000 images will take up around 1GB on the phone.

To buy or to wait?

When shopping for any piece of computer equipment, there's always the tricky question of whether to buy the current model, or hang on for the next version, which may be better *and* less expensive. In the case of Apple products, the situation is worse than normal, because the company is famously secretive about plans to release new or upgraded hardware.

Unless you have a friend who works in Apple HQ, you're unlikely to hear anything from the horse's mouth until the day a new product appears. So, unless a new model came out recently, there's always the possibility that your new purchase will be out of date within a few weeks. About the best you can do is check out sites where rumours of new models are discussed. But don't believe everything you read…

Mac Rumors buyersguide.macrumors.com
Apple Insider appleinsider.com
Think Secret thinksecret.com

Where to buy?

Unlike iPods and Macs, the iPhone is only available direct from Apple or the partner phone carrier in your country. The price will be the same – or at least so close that it's unlikely to be factor.

Buying from a high-street store means you'll get the phone immediately; ordering online you can expect a week's wait for delivery. Another advantage of visiting a store is that you get to see the thing in the flesh and try the various features before you buy. To find your nearest Apple Store, and to check for the availability of iPhones at various different branches, see:

Apple Stores apple.com/retail

Or to find the nearest branch of your carrier, check online, for example, at:

AT&T (US) wireless.att.com/find-a-store/iphone
O2 (UK) www.webmap.o2.co.uk/interfaces/retail

New York City's Fifth Avenue Apple Store

What's in the box?

At the time of writing, the iPhone comes with a stereo headset with mic/button, a USB charging/sync cable, a charger and a polishing cloth.

Used iPhones

Refurbished iPhones

Refurbished Apple products are either end-of-line models or up-to-date ones which have been returned for some reason. They come "as new" – checked, repackaged and with a full warranty – but are reduced in price by anything up to 40 percent. You'll find the Apple Refurb Store on the bottom-right of the Apple Store homepage. If there's something there you want, act quickly, as items are often in short supply.

Secondhand iPhones

It's possible to buy a used iPhone from an individual and get a new SIM card for it from AT&T (in the US) or O2 (in the UK). If the phone has been unlocked, you may be able to use any SIM card and network but be aware that the unlocking will probably have voided the warranty (see p.21).

As with all used electronic equipment, make sure you see it in action before parting with any cash, but remember that this won't tell you everything. If an iPhone has been used a lot, for example, the battery might be on its last legs and soon need replacing, which will add to the cost (see p.89).

Recycling your old phone

You should never throw old mobile phones in the bin. Not only do they contain chemicals that can be harmful to the environment when incinerated or landfilled, they also contain metals and other materials that can be recycled and used in new phones.

If you have any old phones or iPods to dispose of, various groups will take them off your hands – including Apple, which provides addressed, postage-paid envelopes expressly for the purpose. To order one, visit:

Apple Recycling (US) apple.com/recycling
Apple Recycling (UK) apple.com/uk/recycling

Even better, you could donate your old phone to charity – since even broken handsets have some cash value. In the UK, for example, Oxfam can fund 83 school meals in the developing world for each handset dropped off at an Oxfam store or sent to: Oxfam Recycle Scheme, Freepost LON16281, London WC1N 3BR.

Basics

03

Getting started

charging, syncing, typing

Setting up a new iPhone is usually a simple process. With a little luck, and assuming that your computer is up to the job (see p.16), it shouldn't take you more than an hour to get everything up and running and copy across some music, videos, photos, contacts and calendars from your Mac or PC. However, you may first need to grab the latest version of iTunes…

Download the latest iTunes

If you've ever used an iPod, you'll already be familiar with iTunes – Apple's application for managing music, videos and podcasts, ripping CDs and downloading music and video. As with the iPod, iTunes is the bridge between your iPhone and your computer.

If you use iTunes...

Even if you already use iTunes, you may need to update to the latest version to get it to work with the iPhone. New versions come out every month or so, and it's always worth having the latest. To make sure you have the most recent version open iTunes and, on a Mac, choose Check for Updates... from the iTunes menu and, on a PC, look in the Help menu.

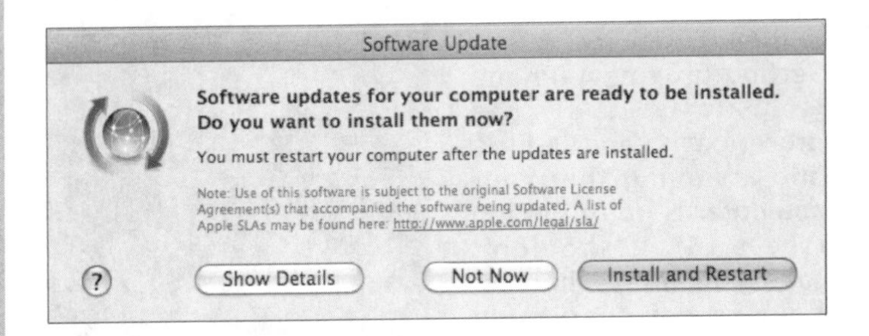

If you don't already use iTunes...

All recent Macs have iTunes pre-installed – you'll find it in the Applications folder and on the Dock. Open it up and check for updates, as described above. If you have a PC, however, you'll need to download iTunes from Apple:

iTunes apple.com/itunes

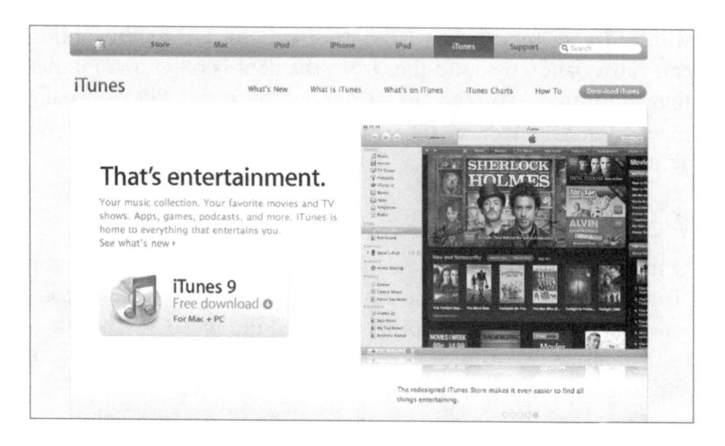

Once you've downloaded the installer file, double-click it and follow the prompts. Either during the installation or the first time you run iTunes, you'll be presented with a couple of questions. Don't worry too much about these as you're just choosing options that can be changed at any time in iTunes Preferences. But it's worth understanding what you're being asked...

• Yes, use iTunes for Internet audio content or
• No, do not alter my Internet settings

This is asking whether you'd like your computer to use iTunes (as opposed to whatever plug-ins you are currently using) as the program to handle sound and files such as MP3s when surfing the web. iTunes can do a pretty good job of dealing with online audio, so in general hitting yes is a good idea, but if you'd rather stick with your existing Internet audio setup, hit No.

• Do you want to search for music files on your computer and copy them to the iTunes Library?

If you have music files scattered around your computer, and you'd like them automatically put in one place, select Yes and iTunes will find and import them all. Otherwise, hit No, as this option can cause random sound files from the depths of your computer to be imported – you can always remove, of course, but it's usually nicer to start with a blank sheet and import only the files you actually want.

Welcome to iTunes

Once everything's up and running you'll be presented with the iTunes window. To the left is the Source List, which contains icons for everything from playlists to connected iPhones. Click any item in the Source List to reveal its contents in the mail section of the window.

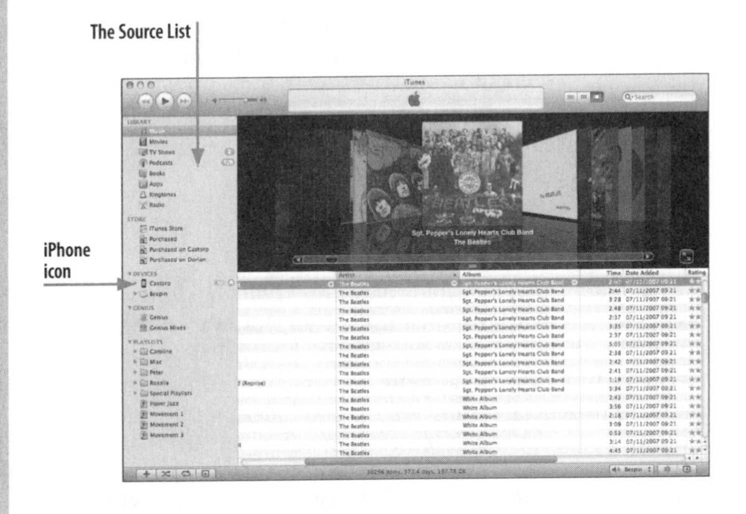

iTunes essential tips

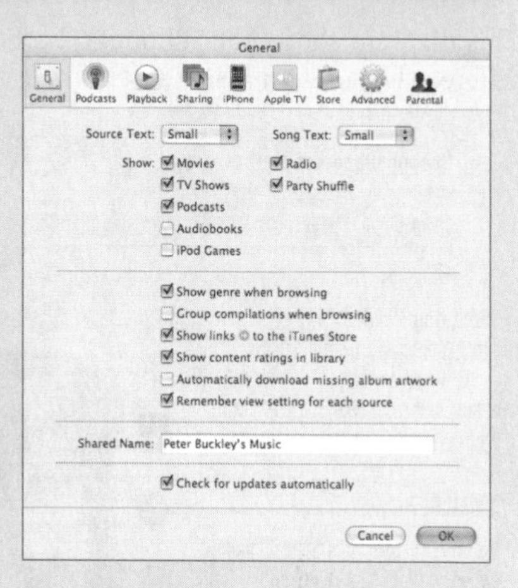

• **iTunes Preferences**, which is referred to throughout this book, contains a whole host of options for changing the way iTunes works. Preferences can be opened via the iTunes menu (Mac) or the Edit menu (PC). Or you could use the following "coma"-key shortcuts:
⌘+, (Mac)
Ctrl+, (PC)

• **Selecting multiple items** To select multiple tracks or other items, hold down the ⌘ key (Mac) or Ctrl key (PC) as you click. Alternatively, to select lots of adjacent tracks at once, click on the first and then Shift+click the last; individual songs can then be removed from this selected group by clicking them while holding down the ⌘ key (Mac) or the Ctrl key (PC).

• **Navigating without the mouse** Once you get the hang of it, whizzing around the iTunes window is much quicker with the keyboard than with the mouse. Use the Tab key to skip between the Source List, the search box and the main panel. To set something playing, hit the Return key.

• **Viewing the total playing time** Just below the main panel, iTunes displays the total amount of playing times and disk space occupied by whatever tracks or videos are visible in the main panel. Note that these are approximations; click the displayed time and it will change to an exact reading.

• **Creating multiple windows** When managing and arranging your music, don't feel obliged to keep everything confined to a single iTunes panel: double-clicking a highlighted playlist's icon or other item icon in the Source List will open its contents in a new floating frame.

For a detailed guide to iTunes, pick up a copy of *The Rough Guide to iPods and iTunes*.

The iPhone at a glance

This diagram shows the iPhone 4. Earlier models have a slightly different layout with a SIM slot at the top.

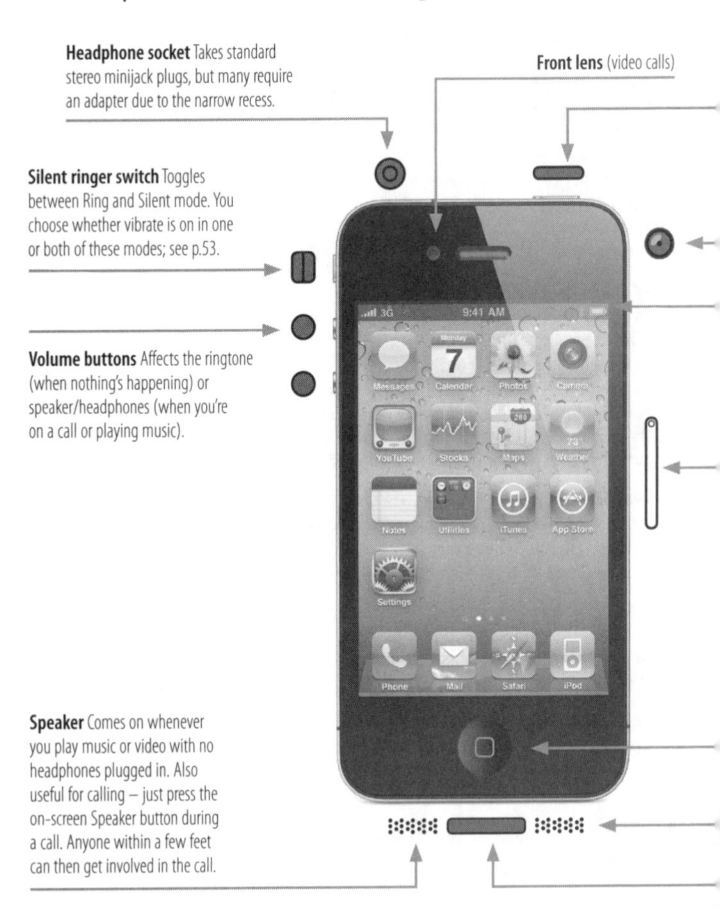

Sleep/Wake Click once to sleep (will still receive calls); hold down for three seconds to power off (will no longer receive calls).

Rear lens (for photos and video)

MicroSIM tray Gives access to the MicroSIM card. To remove it, press the tiny circle with the point of a paperclip.

Status bar Displays the time and gives you feedback about your phone via various icons. These include the following:

- Phone signal level. Relates to calls rather than Internet. When abroad, you'll also see the name of the current carrier.
- Airplane mode on: ie phone, Wi-Fi and Bluetooth signals disabled.
- Phone locked.
- Music or a podcast is currently playing.
- Alarm set. See p.226.
- O E 3G Data network connection – either GPRS (slowest), EDGE (faster) or 3G (fastest), depending on local reception.
- ...which is replaced by this symbol when you're connected to a Wi-Fi network (see p.65). More bars means a stronger signal.
- Bluetooth is on. See p.68.
- Battery charging.
- Battery fully charged.

Home button Click to leave the app you are using and return to the last Home Screen you viewed. Whatever you're currently doing will be put on hold, so you can return to it later. Click again to go to the farthest left Home Screen. Click again to see the Search Screen (see p.57). Double click to see recently used apps (see p.76).

Mic Works well enough from a few feet away when using speaker phone.

Dock connector socket Takes the iPhone sync/charge cable – which is interchangeable with an iPod cable. The socket is also used for certain accessories.

Cables & docks

The iPhone comes with a charge/sync cable just like those used for the iPod. One end can attach directly to the iPhone or – if you choose to buy one – to a little stand known as a "Dock". The other end of the cable connects to any USB port – on a Mac or PC, on the supplied power adapter, or anywhere else.

The Dock

The Dock is a stand that makes it convenient to connect and disconnect the iPhone to computers, power sources, speakers or hi-fis. The Dock's connections can be left in place, so when you get home you simply drop your iPhone in and it's instantly hooked up, synced and charging. In addition, the Dock features a genuine line-out socket, as opposed to a headphone socket, so the sound quality is marginally improved when connecting to a hi-fi or speakers.

Using iPod Docks and cables

If you have a USB cable for an iPod, this should work fine with your iPhone – and vice versa. Some iPod Docks work too, though it depends on the model in question. If you have a Universal Dock, you can buy an inexpensive iPhone adapter for it. One difference between the iPhone Dock and an iPod Dock is that the former offers "special audio porting" – a rather grandiloquent name for some little holes in the base of the cradle that allow you to make use of the iPhone's speaker and microphone while the device is in the Dock.

Charging

To charge an iPhone, simply connect it to a USB port – either on a computer or a USB power adapter. Note, however, that if you're using a computer, the USB port will need to be "powered". Most are, but some, such as those on keyboards and other peripherals, may not work. Also note that your iPhone usually won't charge from a Mac or PC in sleep or standby mode.

When the iPhone is charging, the battery icon at the top-right of the screen will display a lightning slash. When it's fully charged this will change to a plug. When your phone is plugged in and not in use, you'll also see a large battery icon across its centre, which shows how charged it is at present. Like many mobile devices, iPhones use a combination of "fast" and "trickle" charging. This means it should take around two hours to achieve an 80 percent charge, and another two hours to get to 100 percent.

iPhone 4 battery life

Talk time: Up to 7 hours on 3G • Up to 14 hours on 2G

Internet use: Up to 6 hours on 3G • Up to 10 hours on Wi-Fi

Standby time: Up to 300 hours

Video playback: Up to 10 hours

Audio playback: Up to 40 hours

If your iPhone's power becomes so low that the device can't function, you may well find that plugging in to start charging will not revive it straight away. Don't worry – it should come back to life after ten minutes or so.

If you're in a hurry to charge, don't use or sync the iPhone while it's charging – this will slow the charge process. You can cancel a sync with the slider on the iPhone screen – or by "ejecting" the phone in iTunes (see box overleaf). For tips on maximizing your battery life see p.88.

Disconnecting an iPhone

Unlike an iPod, an iPhone doesn't batten down its hatches when connected to your computer. All its functions remain available – you can even make calls. Likewise, whereas an iPod sometimes needs to be properly "ejected" before it can be disconnected, the iPhone can be disconnected at any time, even halfway through a sync. Despite this, iTunes still offers an eject button ⏏ next to the iPhone's icon. This allows you to remove the iPhone from iTunes but leave it charging.

Activating

Once you have the latest version of iTunes installed, and your iPhone is connected to your computer, you're ready to activate the phone and copy across any data and media from your computer. Within iTunes, follow the prompts to register with Apple's partner carrier in your country, transfer your old number and choose a tariff. As part of the process you'll need to enter your iTunes account details (Apple ID or MobileMe details will also work). If you don't have an account, you'll need to set one up – even if you have no intention of buying any music or video via iTunes.

Note that both your activation fee (where applicable) and monthly service-plan payments are billed direct to your carrier – *not* your iTunes account.

Synchronizing

Once the activation process is complete, you'll find yourself presented with the various tabs that control how the iPhone is synchronized with your computer. These include the following, each of which is covered in more detail elsewhere in this book.

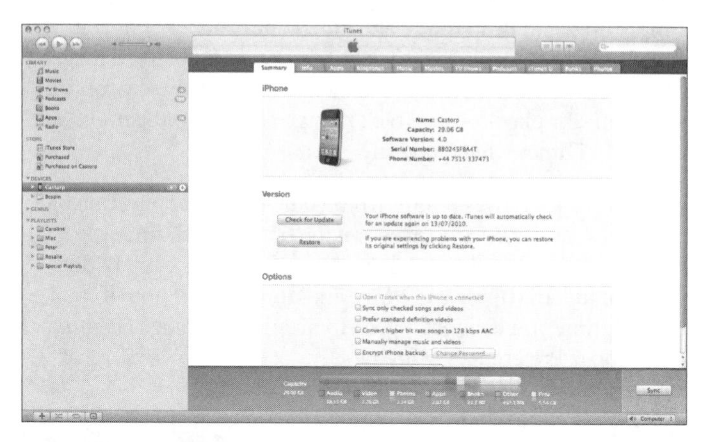

• **Info** Lets you synchronize contacts (see p.102), mail accounts (see p.187), calendars (see p.224) and bookmarks (see p.178) from your computer. Once set up, changes or additions made on the computer will be reflected on your phone, and vice versa.

> **Tip: MobileMe, Google and Exchange users can set up their server-based syncs directly on the iPhone (see p.187).**

• **Apps** Lets you browse all the apps downloaded via your phone or through iTunes. You can choose which to sync over to your phone and even rearrange your apps into iPhone screens and folders – which is often quicker than doing it on the phone itself.

Tip: **If you receive a call while iTunes is performing a sync, the sync will be cancelled to allow you to answer the call. Reconnect the phone after you've hung up, to allow the sync to finish.**

• **Music, podcasts, video & TV shows** Lets you choose which of your iTunes content to sync over to your iPhone. If you'd rather simply drag and drop music and video onto your iPhone, rather than have it synchronized, click Summary > Manually Manage Music and Videos. Note that media downloaded directly onto the phone, or playlists and track ratings created on the move, are copied to iTunes when you sync.

• **Photos** iTunes moves photos from your selected application or folder (see p.206) and gives you the option, each time you connect, to import photos taken with the iPhone's camera onto your computer. (If you use a Mac and don't want iPhoto to pop up each time you connect, open Applications > Image Capture, and choose Preferences from the Image Capture menu. For "When a camera is connected...", select No Application.)

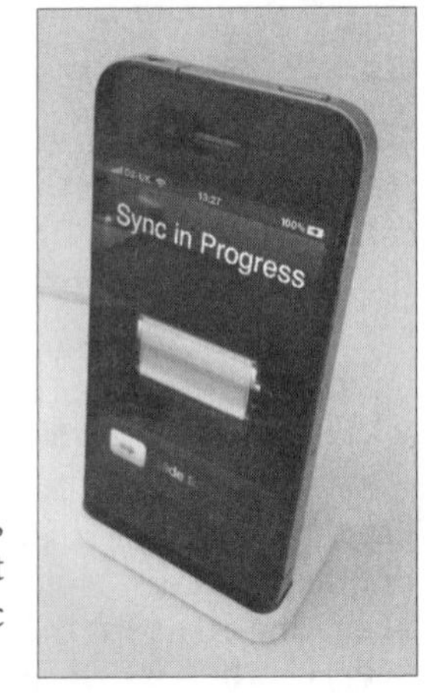

• **Books** This is the tab where you choose how to synchronize PDF and eBook files (the latter in the ePub format) to be read on the iPhone within the iBooks app. To find out more, turn to p.215.

Whenever your phone is connected, you can click the triangle to the left of its icon in iTunes to see what music and other media it's currently storing.

Forcing a sync

When you choose from any of the above options, click Apply Now to start syncing straight away. You can also initiate a sync at any time by right-clicking the icon for your iPhone (Ctrl-click on a Mac) and choosing Sync from the dropdown menu.

Syncing with multiple computers

When you connect your iPhone to a different computer, it will appear in iTunes (as long as it's a recent version of iTunes) with all the sync options unchecked. You can then skip through the various tabs and choose to overwrite some or all of the current content.

- **Music, video & podcasts** Adding music, video or podcasts from a second computer will erase all of the existing media from the phone, since an iPhone can be linked with only one iTunes Library at a time. This applies even if you have Manually Manage Music and Videos turned on. You'll also lose any on-the-go playlists and track ratings entered since your last sync. Next time you connect at home, you can reload your own media, but you won't be able to copy the new material back onto your computer.

> **Tip: You can purchase via multiple iTunes Accounts on an iPhone, but only sync those purchases to iTunes if your computer is "authorized" for the accounts (see p.155).**

- **Apps** You can add apps from a second machine (even if it uses a different iTunes Account) without overwriting existing apps on the iPhone; however, you won't be able to sync all the new apps back to your machine at home without "authorizing" it for the other iTunes Account (see p.152).

• **Photos** can be synced from a new machine without affecting music, video or any other content. However, to leave everything other than photos intact, you need to hit Cancel when iTunes offers to sync the "Account Information" from the new machine.

• **Info** When you add contacts, calendars, email accounts or bookmarks from a second computer, you have two choices – either merging the new data and existing data, or simply overwriting the existing data. iTunes will ask you which way you want to play it when you check the box for a category and click Apply. However, you can bypass this by scrolling down to the bottom of the Info panel and checking the relevant overwrite boxes.

Basic setup options

Once your iPhone is stocked up with all your media and data, you're ready to make it your own.

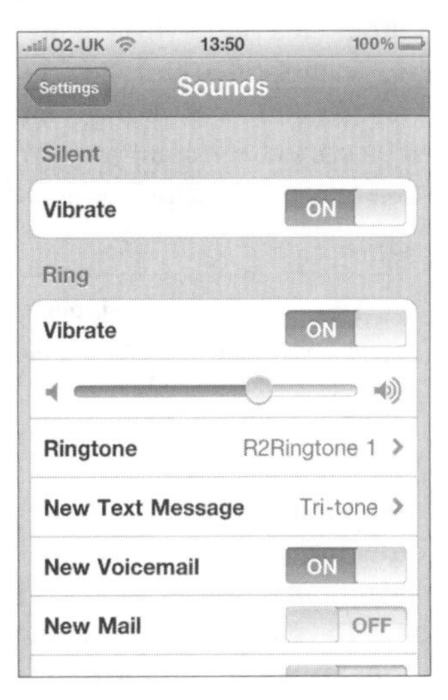

Ringtones and alerts

To audition the built-in ringtones, tap Settings > Sounds > Ringtones. Here you can also toggle various alert sounds on and off, and choose a volume level for your chosen ringtone (this does the same as using the volume buttons on the side of the phone).

Custom ringtones

If the ringtones that come pre-installed on the iPhone aren't to your taste, try adding your own custom ringtones. There are several different ways that this can be done.

The method that Apple would prefer you to use is to buy custom ringtones of your favourite songs from the iTunes Store. It's not free, however: even if you have already purchased the track you have to pay all over again to get the iPhone-compatible ringtone version. What's more, only a fraction of the songs in the store are available as ringtones, and – at the time of writing – the only way to tell whether a track is available as a ringtone, is to view the Store's Ringtone column in iTunes on a Mac or PC when browsing a list of songs. To do this, right-click the header of any column in the list, and choose "Ringtone". For more information, see: apple.com/itunes/store/ringtones.html.

An alternative way to manufacture custom ringtones on a Mac is by using Apple's GarageBand software. First, select a loop of under 40 seconds (using the Cycle button) from either your own composition or an existing song file that you've imported. Then choose Send Ringtone to iTunes… from the Share menu.

Whether purchased or custom-made, your ringtones end up in your iTunes Library, where they can be found by clicking Ringtones in the sidebar. Next time you connect your iPhone, click the Ringtones tab and check the box to sync your tones across to the phone.

There are also several third-party applications that allow you to create and sync custom ringtones to your iPhone, for free, independently of iTunes. iToner, for example, has a very easy-to-use floating iPhone-like interface that let's you convert entire iTunes playlists into ringtones on-the-fly and then deposit them onto your iPhone.

iToner (Mac) ambrosiasw.com/utilities/itoner
iRingtoner (PC) tinyurl.com/iphonerlink

Ring, vibrate and silent

The iPhone offers two modes – Ring and Silent – which you can select using the switch on the left-hand side of the phone. You can choose to have the phone vibrate in one or both of these modes. Click Settings > Sounds to make your selection.

Wallpaper

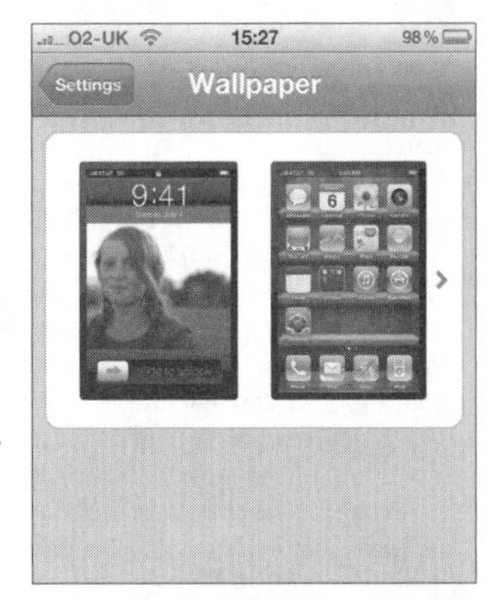

Tap Settings > Wallpaper to choose the photos that appear on your Lock Screen (the one you see when you wake the iPhone from sleep mode) and your Home Screen (the screen on which all your icons sit). You can drag and crop the photo before setting it. When setting a Lock Screen image, set the top and bottom edges of a horizontal photo to align with the edges of the semi-transparent zones; the pic will appear framed by, rather than behind, the strips when the phone awakes from sleep. The Home Screen image needs a little more consideration, as a fussy image can make your iPhone pretty ugly and hard to use. The plainer the better … or you could try something custom, such as that pictured here (found at tinyurl.com/iPh4Wall), which mimics the iBooks app bookshelf.

iPhone name

You will probably also want to change the unimaginative name that iTunes gives your iPhone during activation. To do this, tap on the name next to the iPhone icon in iTunes and retype whatever you want.

Auto-Lock

Within Settings > General, you can set the number of minutes of inactivity before your iPhone goes to sleep and locks its screen. In order to maximize battery life, leave it set to 1 minute unless you find this setting annoying.

Passcode Lock

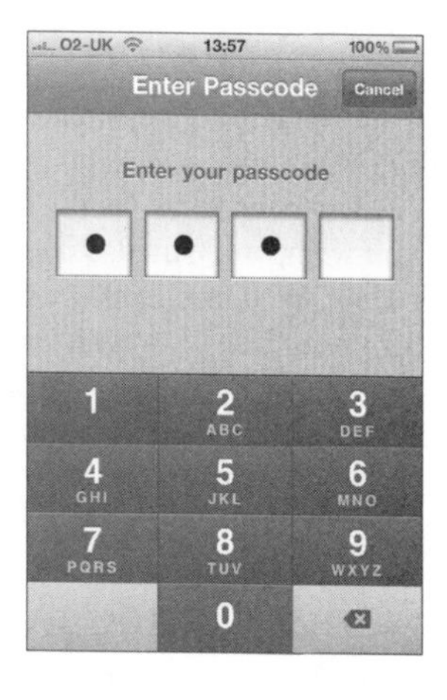

If you want to protect the private data on your phone – and make sure no one ever uses it to make calls without your permission – apply a passcode. Tap Settings > General > Passcode Lock, and choose a 4-digit number. With this feature turned on, the phone can't be unlocked after waking from sleep without first entering the number. If you forget the code, connect your phone to your computer and restore it (see p.84). If a thief tries this, they'll get the phone working, but by that stage all your private data will no longer be on the phone. Here, you can also turn on the Erase Data setting, so that after ten failed passcode attempts, your iPhone wipes itself.

If you would rather use a proper alphanumeric password instead of a 4-digit pin, slide the Simple Passcode setting off and enter something more complicated.

> **Tip: Another less serious privacy risk is that, when you receive an SMS, the message will pop up on the screen even if the phone is locked. If you'd rather this didn't happen, turn off SMS Preview within the Passcode Lock page.**

Parental controls

Any iPhone running software version 2.0 or later also features parental controls, branded Restrictions. Locate the relevant panel within Settings and you'll find a host of ways to make the phone child-friendly – blocking access to YouTube, for example, or to iTunes Store and App Store content tagged as "explicit". If you wish, you can also use this panel to completely restrict app purchases, or even use of the iPhones camera, FaceTime and location services … but you'd have to be a pretty mean parent to turn them all off.

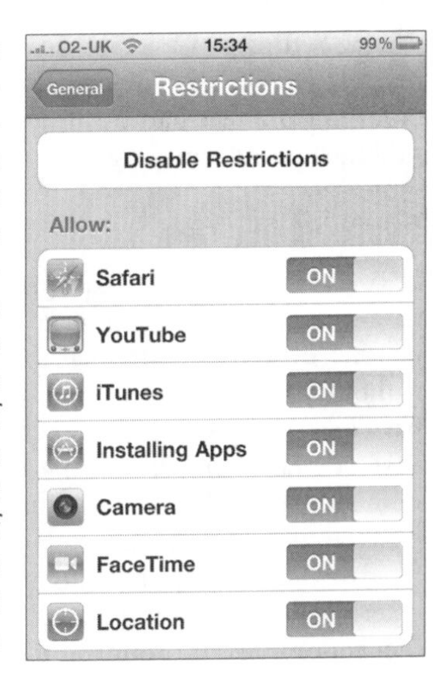

Accessibility tools

The iPhone comes loaded with several accessibility tools designed to aid the blind and visually impaired. To see what's available, and play with the options, tap through to Settings > General > Accessibility. Features include:

- **VoiceOver** Used to speak items of text on the screen. Touch an item to hear it spoken, double-tap to select it and use three fingers to scroll.

- **Screen zoom** Once set up, double-tapping with three fingers will zoom you in and out of the screen; dragging with three fingers moves you around the screen.

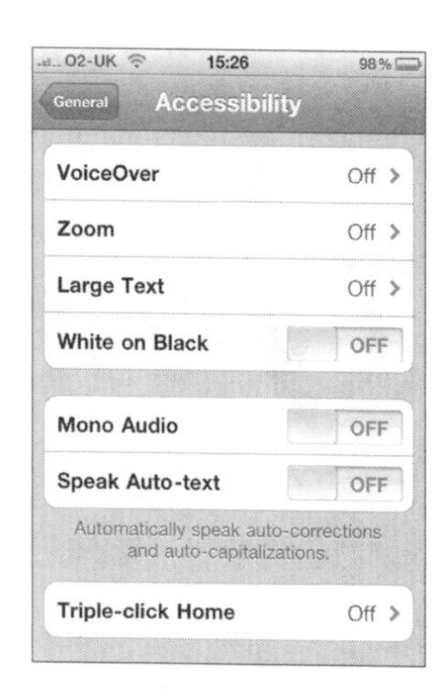

• **Large Text** Does what it says on the tin for text in the Contacts, Mail, Messages and the Notes applications; there are several different sizes to choose from.

• **White on Black** This setting renders the screen as a negative image, making blacks white, whites black, etc.

• **Audio settings** Aside from enabling mono audio for the headphone socket of your iPhone, you can also set the phone to automatically speak auto-corrections and auto-capitalizations.

• **Video settings** For those whose hearing is impaired, an option for turning on closed-captioning for video can be found within Settings > iPod.

Search settings

When viewing the farthest Home Screen to the left, the iPhone's Search Screen is accessed by either a swipe to the left, or a single tap of the Home button.

Get into the habit of quickly popping to the Search Screen to find emails, calendar events or a friend's contact details. You'll soon find that it's much faster than navigating via specific apps.

To choose exactly what content from your iPhone is searched, check and uncheck the options within Settings > General > Spotlight Search. Here, you can also change the order in which results are displayed, by dragging the ≡ icons up and down.

04

Typing tips

How to get the best from the iPhone's virtual keyboard

The iPhone's touch-screen keyboard isn't to everyone's taste, but if you can get used to it, the iPhone allows for typing far faster than almost any of its competitors. Following are some tips to get you started. The best place to practise is in the built-in Notes application.

Basic techniques

- **Numbers and punctuation** To reveal these keys, tap .?123.

- **Symbols** To reveal these keys, tap .?123 followed by #+=.

- **Edits and navigating** You can tap anywhere in your text to jump to that point. For more accuracy, tap, hold and then slide around to see a magnifying glass containing a cursor.

- **Pop-up keys** The iPhone enters a letter or symbol when you release your finger, not when you touch the screen. So if you're struggling to type accurately (or you're entering a password and can't see what you're typing) try tapping and holding the letter. If the wrong letter pops up, slide to the correct one and then release. Also note that many keys, when tapped and held, reveal alternative characters for you to slide your finger over and select.

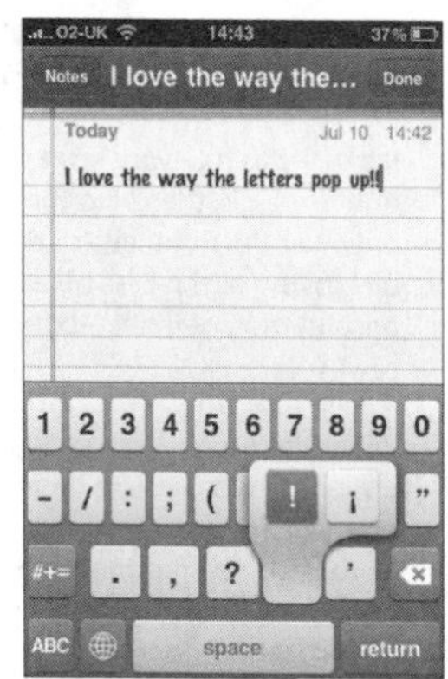

Speed-typing tips

• **Quick full stops** When you reach the end of a sentence, double-tap the space bar to add a full stop and a space. If this trick doesn't work, turn the relevant option on in Settings > General > Keyboard.

• **One-touch punctuation** Tap .?123 and then slide to the relevant key without taking your finger off the screen. Much more convenient than tapping twice.

• **One-touch caps** The same trick works with capital letters: tap shift and slide to a letter.

• **Caps Lock** If you like to TYPE IN CAPS, turn on the Caps Lock feature under Settings > General > Keyboard. Once done, you can double-tap the Shift key (⇧) to toggle Caps Lock on and off.

• **Thumbs and fingers** Apple recommend that you start off using just your index finger and progress to two thumbs. However, you could also try multiple fingers on one hand. This can be even faster, since thumbs tend to bump into each other when aiming at keys near the middle of the keyboard, though it does take a little getting used to.

Auto-correct and spell check

It's tricky to hit every key accurately on the iPhone, but usually that doesn't matter much, thanks to the device's word-prediction software. Even if you hit only half the right letters, the phone will usually work out what you meant by looking at the keys adjacent to the ones you tapped and comparing each permutation of letters to those in its dictionary of words.

Accepting and rejecting suggestions

When the iPhone suggests a word or name it will appear in a little bubble under the word you're typing. To accept the suggestion, just keep typing as normal – hit space, return or a punctuation mark. To reject it, finish typing the word and then tap the suggestion bubble before continuing.

Dictionary

The iPhone has a much bigger and more relevant dictionary than most mobile phones – including, for example, many names and swear words. In addition, it learns all names stored in your contacts and any word that you've typed twice and for which you've rejected the suggested correction. Unfortunately, it's not currently possible to edit the dictionary, but you can blank it and start again. Tap Settings > General > Reset > Reset Keyboard Dictionary.

If you're ever stuck for the spelling or meaning of an obscure word, try a dictionary web app such as idotg.com/apps, or one of the many downloadable dictionary apps available in the App Store.

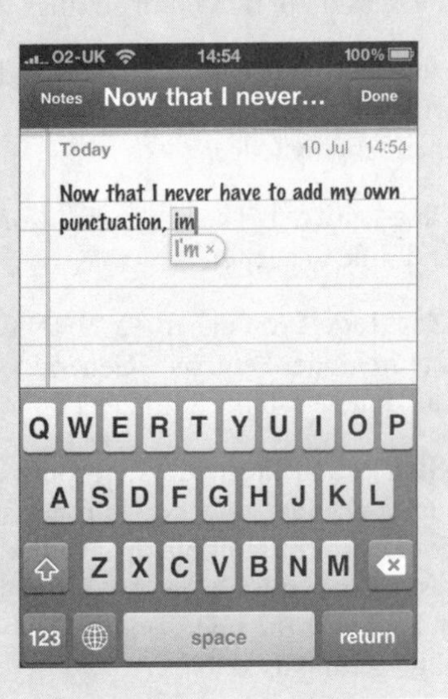

Spell check

After you've typed a word that the iPhone doesn't recognize, it will be underlined in red to suggest that it might be a spelling error. If you want to check, tap the word to be presented with an alternative spelling from the dictionary.

• **One touch at a time** If typing with two thumbs or multiple fingers, only let one finger touch the screen at a time. If the first finger is still on the screen, the second tap won't be recognized.

• **Auto-capitalization** In addition to correcting letters, the iPhone will add punctuation (changing "Im" to "I'm", for example) and capitalize the first letter of words at the start of sentences. If you prefer to stick with lower case, turn off Auto-capitalization within Settings > General > Keyboard.

• **Landscape keyboard** In Safari, Mail and many other applications, you can rotate the iPhone to get a bigger version of the keyboard.

• **Test your skills** There are loads of apps available in the App Store that you can use to test your typing speed, and also train yourself to get faster. TypeFast is a good choice, as is Typing Class, which doubles as a game, and is great for kids. There are also several iPhone formatted websites that can help, such as iphone.typingweb.com.

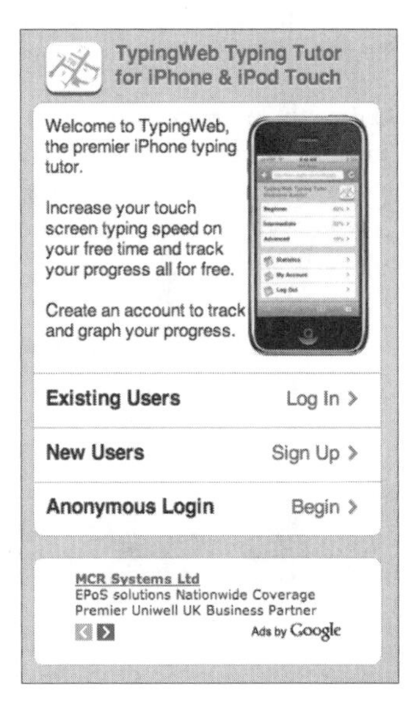

Cut, copy & paste

In terms of typing, among the most useful features on the iPhone are the Cut, Copy and Paste commands – especially if you don't particularly get on with the device's keyboard. Though at first they can seem a little fiddly to use, it really is worth getting acquainted with the taps and drags that bring these controls to life. And we are not only talking about text; these options will also pop up for copying images, and even entire webpages. For now, however, here's the deal with text:

• **1** Locate the text or word you want to copy or cut and then tap and hold until an options bubble appears; choose Select to highlight that word, or Select All for the whole piece of prose of which your word is a part. To speed things up, you could also try tapping once at the start of the text you want to select and then immediately dragging to the end last word.

• **2** Use the blue end-point markers to resize the selection you have highlighted, and then when you are ready, tap Copy or Cut in the option bubble. Both actions move the selection onto the iPhone's "clipboard", ready to be pasted elsewhere.

> **Tip:** **If you want a specific app for organizing copied content on a clipboard, download the excellent Pastebot.**

• **3** Navigate to the place where you want to paste the text (which might be a completely different app, an email you are composing, or perhaps the search field in Safari).

• **4** Tap and hold at your desired point and choose Paste from the option bubble that appears. You can alternatively Select text (as above) and then Paste over the top of it.

> **Tip:** When copying, pasting and typing, you can undo your last action by shaking the iPhone.

Replace

The option bubble that appears when you select composed text sometimes also offers up the Replace command. This has nothing to do with replacing text with the contents of the clipboard – that's what Paste is for – but is instead used to find common alternatives to words that you may have mistyped. It's not one hundred percent accurate (as pictured below), but can still be a useful alternative to retyping the whole word.

> **Tip:** If you fancy something a bit old school, try a virtual old-fashioned typewriter app, such as miTypewriter.

05

Connecting

Wi-Fi, Bluetooth and other airwaves

The iPhone can handle various kinds of wireless signal: GSM, for phone calls; GPRS, EDGE and 3G for mobile Internet access; Wi-Fi, for Internet access in homes, offices and public hotspots; and Bluetooth, for connecting to compatible headsets and carphone systems. This chapter takes a quick look at each.

Airplane mode

Airplane mode, quickly accessible at the top of the Settings menu, lets you temporarily disable GSM, GPRS, EDGE, 3G, Wi-Fi and Bluetooth, enabling you to use non-wireless features such as the iPod app during a flight or in any other circumstances where mobile phone use is not permitted. (Whether phones actually cause any risk on aircraft is disputed, but that's another story.) Airplane mode can also be useful if you want to make sure you don't incur roaming charges when overseas by inadvertently checking your mail and so on.

Using Wi-Fi

Connecting to networks

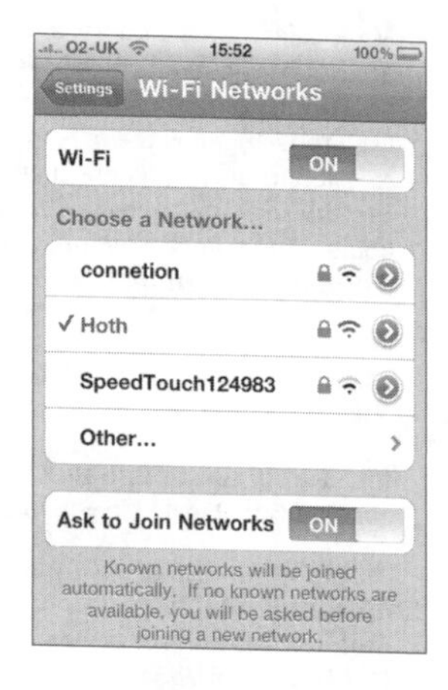

To connect to a Wi-Fi network, tap Settings > Wi-Fi and choose a network from the list. If it's a secure wireless network (as indicated by the padlock icon), the iPhone will invite you to enter the relevant password.

Though this procedure doesn't take long, it's best to have the iPhone point you in the direction of Wi-Fi networks automatically. This way, whenever you open an Internet-based tool such as Maps or Mail, and there are no known networks in range, the iPhone will automatically present you with a list of all the networks it can find. You can turn this feature on and off via Settings > Wi-Fi > Ask to Join Networks.

If the network you want to connect to isn't in the list, you could be out of range, or it could be that it's a "hidden" network, in which case tap Wi-Fi > Other and enter its name, password and password type.

Tip: **If you want to maximize battery life, get into the habit of turning Wi-Fi off when you're not using it. It only takes a couple of seconds to turn it back on when you next need it.**

Forgetting networks

Once you've connected to a Wi-Fi network, your iPhone will remember it as a trusted network and connect to it automatically

Finding public hotspots

Many cafés, hotels, airports and other public places offer free wireless Internet access, though often you'll have to pay for the privilege of using them – particularly in establishments which are part of big chains (Starbucks and the like). You either pay the person running the system (over the counter in a café, for instance) or connect and sign up on-screen. If you use such services a lot, you may save time and money by signing up with a service such as Boingo, T-Mobile or Wayport, which allow you to connect at thousands of hotspots for a monthly fee.

Boingo boingo.com
T-Mobile t-mobile.com/hotspot
Wayport wayport.com

The ideal, of course, is to stick to free hotspots. There are various online directories that will help you locate them, though none is comprehensive:

Hotspot Locations hotspot-locations.com
WiFinder wifinder.com
Wi-Fi Free Spot wififreespot.com
ZONE Finder wi-fi.jiwire.com

whenever you're in range. This is useful, though can be annoying – if, for example, it keeps connecting to a network you once chose accidentally, or one which lets you connect but doesn't provide web access. In these cases, click on the ⊙ icon next to the relevant network name and tap Forget This Network. This won't stop you connecting to it manually in the future.

When it won't connect...

If your iPhone refuses to connect to a Wi-Fi network, try again, in case you mistyped the password or tapped the wrong network name. If you still have no luck, try the following:

• **Try WEP Hex** If there's a ⊙ icon in the password box, tap it, choose WEP Hex and try again.

• **Check the settings** Some networks, especially in offices, require you to manually enter information such as an IP address. Ask your network administrator for the details and plug them in by clicking ⊙ next to the relevant network name.

• **Add your MAC address** Some routers in homes and offices (but not in public hotspots) will only allow access to devices specified in the router's "access list". If this is the case, you'll need to enter the phone's MAC address – which you'll find within Settings > General > About > Wi-Fi Address – to your router's list. This usually means entering the router's setup screen and looking for something titled MAC Filtering or Access List.

• **Reboot the router** If you're at home, try rebooting your wireless router by turning it off or unplugging it for a few seconds. Turn off the Wi-Fi on the phone (Settings > Wi-Fi) until the router has rebooted.

• **Tweak your router settings** If the above doesn't work, try temporarily turning off your router's wireless password to see whether that fixes the problem. If it does, try choosing a different type of password (WEP rather than WPA, for example). If that doesn't help, you could try updating the firmware (internal software) of the router, in case the current version isn't compatible with the iPhone's hardware. Check the manufacturer's website to see if a firmware update is available.

Test your speed

To test the speed and latency (time lag) of your current EDGE, 3G or Wi-Fi signal, try a free app such as Speedtest, which you can use to keep records of your connection speeds from day to day; alternatively, tap Safari and visit:

iPhone Speed Test testmyiphone.com

GSM, GPRS, EDGE & 3G

In your home country, the iPhone should automatically connect to your carrier's GSM network for voice calls, and the fastest data network available – either GPRS, EDGE or 3G. All three of the date networks will automatically give way to Wi-Fi (which is usually much faster) whenever possible.

> **Tip: If you have an iPhone with 3G and want to maximize battery life, consider turning the 3G connection off within Settings > Network. Once that's done, your phone will use EDGE instead, which is slower but far less power-hungry.**

Connecting abroad

When overseas, voice calls should work normally, though North Americans may first need to activate international roaming with AT&T. You'll probably find that your phone selects foreign carriers automatically. If you prefer, however, it's possible to specify a preference. Simply tap Settings > Carrier and pick from the list. For more on making calls abroad, see p.22.

As for Internet, Wi-Fi should work wherever you are in the world – for free, if you can find a non-charging hotspot. For mobile Internet, GPRS, EDGE or 3G will work if there's a compatible network and you've turned on Data Roaming within Settings > General > Network. But be prepared for some extremely steep usage charges (see p.22).

Bluetooth

Bluetooth allows computers, phones, printers and other devices to communicate at high speeds over short distances. As already discussed, however, Apple have disabled all Bluetooth activity on the iPhone except connecting to mono headsets and certain carphone systems. Other devices may recognize the presence of an iPhone, but they can't communicate with it in any useful way.

You can turn Bluetooth on and off by tapping Settings > General > Bluetooth. If you're not using it, leave it switched off to help maximize battery life.

Connecting to office networks

Exchange – for calendars, contacts, etc

Most offices run their email, contacts and calendars via a system known as Microsoft Exchange Server. Individual workers access these various tools using Outlook, and a web address is usually made available to allow remote log-in from home or elsewhere.

If necessary, it's usually possible to access your work account via the web on the iPhone – just press Safari and go to the regular remote-access web address. However, it's much neater to point your phone at the Microsoft Exchange Servers directly. This allows you to access office information via the iPhone's Mail, Calendar and Contacts features, rather than going through the fiddly process of logging in via Safari.

All you need to do is set up your work account as a new email address (see p.187), and then turn calendars, contacts and tasks on or off within the Settings > Email screen. If it doesn't work, speak to your office network administrator for advice.

VPN access

A VPN, or virtual private network, allows a private office network to be made available over the Internet. If your office network uses a VPN to allow access to an Intranet, file servers or whatever, you'll probably find that your iPhone can connect to it. The phone supports most VPN systems (specifically, those which use L2TP or PPTP protocols), so ask your administrator for details and enter them under Settings > General > Network > VPN.

Remote access

As well as connecting via a VPN, it's also possible for the iPhone to connect directly to a computer that's on the Internet – a Mac or PC in your home or office, say. Once set up, you could, for example, stream music and video from your iTunes collection or browse your files. All you need is the right app and the correct security settings set up on the computer in question.

Some of the various apps available for this purpose – all of them available from the App Store – include NetPortal (pictured), LogMeIn Ignition and TeamViewer.

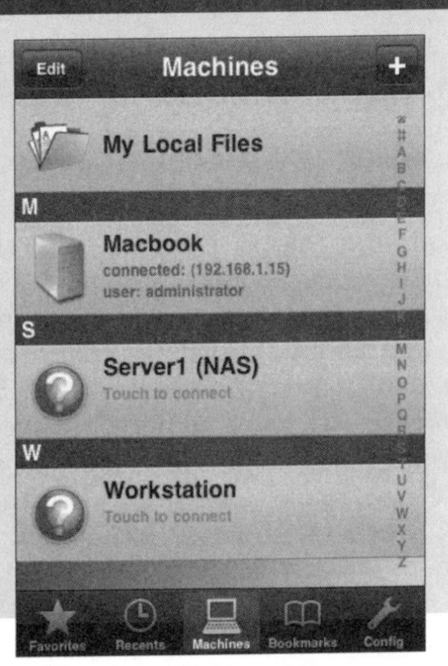

06

Apps & webclips

downloading, creating & organizing apps

Perhaps the single best thing about the iPhone is the sheer number of applications – or apps – available for it. Tens of thousands of apps are available, the vast majority of which can be downloaded either freely or inexpensively. An app might be anything from a driving game to metro map, from a tool for making calls across the Internet to a version of your favourite website optimized to work perfectly on the iPhone's small screen. If you can imagine an application for your iPhone, someone has probably already created it and made it available.

The iPhone comes pre-installed with a bunch of apps – from Mail and Safari to Weather and Stocks. But these are only the tip of the iceberg. To see what else is available, dive into the App Store…

Downloading apps

Just as the iTunes Music Store changed music-buying habits, Apple's App Store – which provides apps for the iPhone, the iPod Touch and the iPad – is quickly changing the way in which software is distributed. With so many apps available at such little cost, the only problem is the potential for getting somewhat addicted.

The App Store can be accessed in two ways:

• **On the iPhone** Simply click the App Store icon. Assuming you're online, you can either browse by category, search for something you know (or hope) exists, or take a look at what's new, popular or featured. When you find something you want, hit its price tag and follow the prompts to set it downloading. You'll need to enter your iTunes or Apple ID that, hopefully, you will have set up when you activated your iPhone.

Apps can be downloaded via 3G, EDGE or Wi-Fi, though some larger apps can take an eternity with anything less than a Wi-Fi connection.

• **On a Mac or PC** Open iTunes, click the Store icon and then hit App Store in the top menu. All the same apps are available and the interface for browsing them is, if anything, better than the one on the phone itself. Any apps you download in iTunes will be synced across to your iPhone next time you connect.

Different countries have their own iTunes Store, and you can only download apps from the one in the region where your iTunes

iPhone & iPad apps

If you own both an iPad and an iPhone, it's worth noting that there are many apps available in the Store that have been specifically designed to run on both iPads and iPhones, but with very different user interfaces. These apps might, for example, offer an elegant split-screen interface on the iPad, and a more stripped-down interface on the iPhone, and are listed within the App Store displaying a small white + icon next to their price, and you only need to buy it once to have it working on both devices.

Account is registered. Note that there are no refunds in the App Store, so it pays to read reviews before you buy.

iTunes Store Accounts

You can log into the App Store using the same username and password that you use to purchase music, books or videos in iTunes or the iBookstore. If you don't already have an account set up, try to download an app in the store and you will quickly be prompted to do so.

What many people don't realize is that neither the iPhone, nor iTunes on your computer, have to be wedded to a single iTunes Account. So if more than one member of your household uses your iPhone (or iTunes on your Mac or PC), there's no reason why you can't all have your own iTunes Store Accounts and buy apps separately. What's more, once apps have been installed on the iPhone they are available to use, whichever account is currently logged in to the iPhone's Store.

It only becomes an issue when you try to copy apps back to iTunes on a Mac or PC, as this will only work if the computer in question is "authorized" for the Store Account that was used to download the apps. Up to five computers can be authorized for each Store Account. To authorize iTunes for a new account, select Store > Authorize Computer and follow the prompts.

You can add apps to multiple iPads, iPhones and iPods from iTunes without having to authorize them or overwrite any apps that might already be installed.

> **Tip:** To log out of your account on the iPhone, either tap Settings > Store > Sign Out, or scroll to the bottom of most App Store windows and tap your Account: name button.

Syncing your downloaded apps

Apps downloaded via iTunes are automatically synced to the iPhone next time you connect. But if you want to be more selective about what does and doesn't sync, connect to iTunes, highlight your iPhone in the sidebar and then make your selections by checking the boxes under the Apps tab.

Apps purchased on the iPhone are automatically synced back to iTunes when you connect, assuming that the Sync Apps box has

Checking app settings

Just like the apps that come pre-installed on the iPhone, many third-party apps have various preferences and settings available. Many people overlook such options and might end up missing out on certain features as a result. Each app does its own thing, but expect to find something either:

- **In the app** If there is nothing obvious labelled Options or Settings, look for a cog icon, or perhaps something buried within a More menu.
- **In Settings** Tap Settings on the Home Screen and scroll down to see if your app has a listing in the lower section of the screen. Tap it to see what options are available.

been highlighted under the Apps tab in iTunes and your computer has been authorized (see p.73) for the iTunes Account used to purchase the apps.

Updating apps

One of the best features of the App Store is that as and when developers release updates for their software, you will automatically be informed of the update and given the option to install it for free, even if you had to shell out for the original download. The number of available updates at any one time is displayed within a red badge on the corner of the App Store's Home Screen icon. To update apps:

• **On the iPhone** Choose App Store > Updates and then Update All. Updates are then synced back to iTunes next time you connect.

> **Tip: You can also selectively choose which updates you download by tapping the individual FREE buttons.**

• **On a Mac or PC** Highlight Applications in the sidebar (where a badge displays the number of available updates) to view all your apps and then click Check for updates, bottom-right, to see what's new (pictured here).

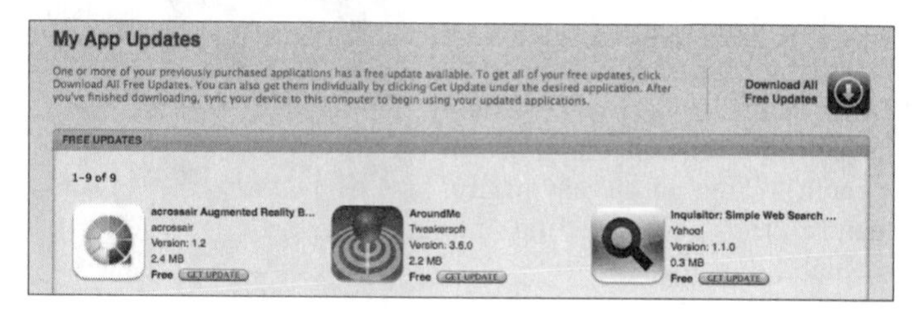

Running and switching apps

To launch an app on your iPhone, simply tap its icon. To close it, press the iPhone's Home button – though as we'll see, the app may continue to run in the background..

Fast app switching

If you own either an iPhone 4 or iPhone 3GS, then double-clicking your phone's Home button reveals your most recently used applications, enabling you to quickly switch between them without having to return to the Home Screen.

> **Tip:** **If you own an older iPhone, double-tapping the Home button generally reveals the iPod controls or takes you to your call Favorites screen. Look for the option within Settings > General > Home.**

Swiping to the left across this tray reveals the next four most recently used apps, and so on. Tapping an app's icon from this tray swiftly switches you into that app, and for many apps you will find that you are back in exactly the same place you were last time you used it (in contrast to having tapped the same app's Home Screen icon, which would have rebooted the app from scratch). This is a very useful feature, and worth getting into the habit of using.

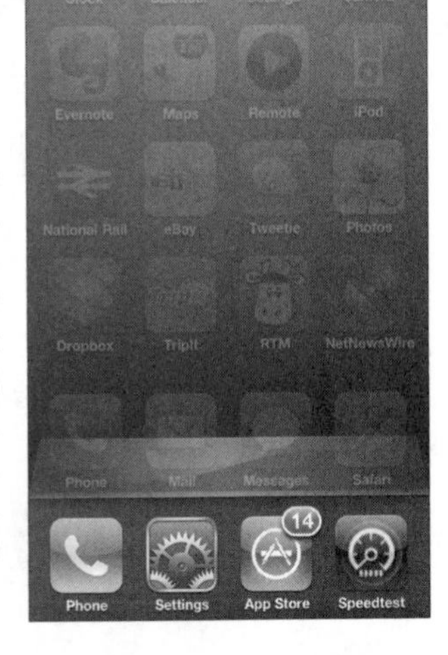

Should you want to remove an app from this tray, tap and hold any of the tray icons until they all start to jiggle about; then tap the red ⊖ icon of the apps you want rid of. Click the Home button when you are done. This process does not remove the app from the iPhone all together, only from the tray.

Switching to the iPod app

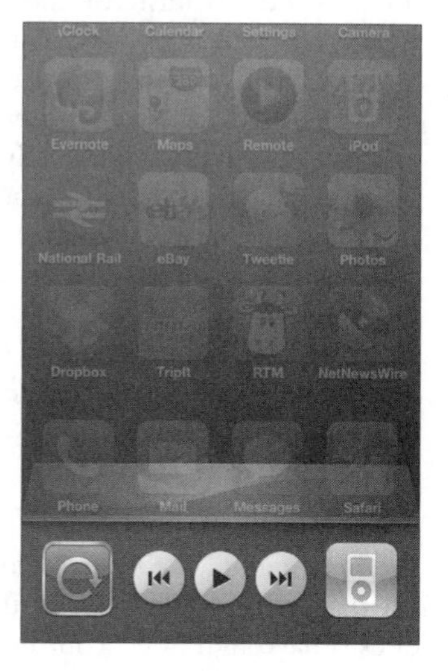

Swiping to the right across the Recently Used App Tray reveals the iPod play controls and an icon that gives you quick access to the iPod app.

To the left of these playback controls there's a special button for locking the iPhone's screen orientation to portrait (handy when trying to use an iPhone reader app in bed). When the screen orientation is locked, this button displays a padlock in its centre.

Multitasking

The iPhone's multitasking capability allows apps to perform certain tasks in the background, while you get on with something else. Check out the specific details for different apps in the App Store, but the kind of functions that can run in the background include music playback (great for radio tuner apps), receiving VoIP calls (for more VoIP services, such as Skype, see p.125), and location awareness (allowing apps to dish up location-specific info or GPS directions from the background).

Creating webclips

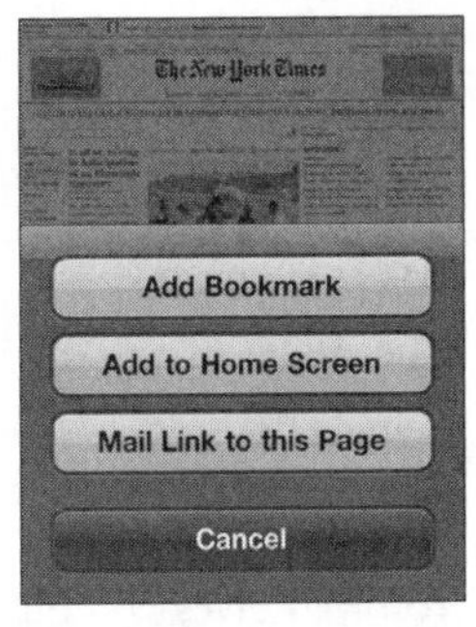

Downloading apps is one way to fill up your Home Screen with icons; the other is to create simple bookmark icons that point to your favourite websites or "web apps" (see box, opposite). This can be handy as it allows you to access your most frequently viewed websites without opening up Safari and manually tapping in an address or searching through your bookmarks.

To create an icon for a website, simply visit the page in Safari and press **+**. Select "Add to Home Screen" and choose a name for the icon (the shorter the better, as anything longer than around ten characters won't display completely on the Home Screen).

Improving website icons

When you add a webpage to your Home Screen, the iPhone will use that website's iPhone icon, if one has been specified. If the site doesn't have an iPhone icon, an icon will be created based on how the page was being displayed when you clicked **+**. To make the best-looking and most readable icons, try zooming in on the logo of the website in question.

If you're not happy with the icon you get, it is possible – if a bit of a hassle – to create your own icons for those websites. The process involves creating an icon (or grabbing one from elsewhere), uploading it to the web, and then using a special bookmark to point the iPhone to this icon before you hit **+**. To grab the special bookmark and get more details, see:

AllInTheHead.com tinyurl.com/addiphoneicons

"Web apps" and optimized sites

Whereas a "proper" iPhone app is downloaded to your phone and runs as a standalone piece of software, a "web app" is an application that takes the form of an interactive webpage. The term is also often used to describe plain old webpages specially designed to fit on the iPhone screen without the need for zooming in and out, as such, you might also hear people refer to them as iPhone-optimized webpages and websites.

Unlike many "proper" apps, for most web apps to work at all you will need to be online. That said, there are some web apps (such as Google's Gmail) that employ "databases", also called "super cookies" (see p.180), to enable some functions to be carried out even when the iPhone can't get online. On the plus side, web apps take up no space in your iPhone's memory, offer near-unlimited possibilities (everything from calculating tips at restaurants to playing chess is covered) and are easily shared with friends (as all you need to do is pass on a web link).

Several websites offer directories of iPhone-optimized websites and web apps, arranged into categories. These can be useful to locate tools when you're out and about, or just fun to browse. Apple's own is very comprehensive, but not especially easy to use on the iPhone as it displays in the form of a regular webpage.

Apple's web app directory apple.com/webapps

Whether or not they call them "web apps", many popular websites offer a version for browsing on mobile devices in general or iPhones specifically. These not only fit nicely on small screens but also load much faster, since they come without the large graphics and other bells and whistles.

Sometimes, if the server recognizes that you're using a mobile web browser, the mobile version of a site will appear automatically. At other times, you'll have to navigate to the mobile version manually – look for a link on the homepage. Mobile sites often have the same address but with "m", "mobile" or "iphone" after a slash or in place of the "www". For example:

Digg digg.com/iphone
eBay mobile.ebay.co.uk
Facebook iphone.facebook.com
Flickr m.flickr.com
MySpace m.myspace.com

Organizing icons

As your iPhone fills up with apps and webclips, they will soon start to spill over onto multiple screens, which you can easily switch between with the flick of a finger – simply slide left or right.

The number of screens currently in use is shown by the row of dots along the bottom of the screen, with the dot representing the screen currently being viewed highlighted in white. You can return to the Home Screen by pressing the Home button or by sliding left until you reach it.

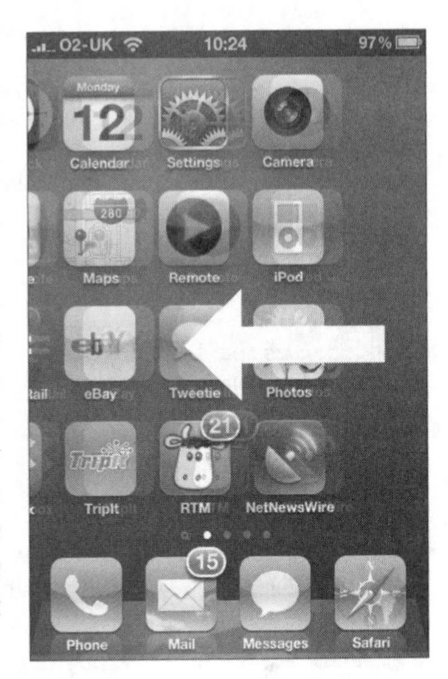

Moving icons

To rearrange the apps and webclips on your iPhone, simply touch any icon for a few seconds until all the icons start to wobble. You can now drag any icon into a new position, including onto the "Dock" at the bottom of the screen (to put a new icon here, first drag one of the existing four out of the way to clear a space).

To drag an icon onto a different screen, drag it to the right- or left-hand side of the screen. (To create a new screen, simply drag an icon to the right-hand edge of the last existing screen.)

Once everything is laid out how you want it, click the Home button to fix everything in place.

Creating folders

Instead of just having your apps and links arranged across various screens, it's also possible to group relevant icons together into folders. You might create, for example, a "travel" folder, which contains a mix of map-based apps, guidebooks and links to travel-related websites.

You could simply group related icons together onto specific screens, but folders offer some advantages. For example, they can be titled and live on your Home Screen, which makes them instantly available at all times.

To create a folder, first touch any icon for a few seconds until they start wobbling. Then simply drag one icon onto another and a folder will be created containing those two icons. The iPhone may attempt to give your new folder a name based on the categories of apps that you've combined – tap into the title field to overwrite this with a name of your choice. To add another app to the folder, simply drag it onto the folder's icon. To remove one, click on the folder and simply drag the relevant icon out.

Tip: Once you have lots of icons, you may find it easier to organize them into screens and folders in iTunes, which allows you to see multiple screens at once and to move icons around with a mouse. Simply attach your iPhone to your computer, select its icon in iTunes and click the Apps tab.

Deleting apps and webclips

It's worth noting that your iTunes Account keeps a permanent record of which apps you have downloaded, so if you do delete both your iPhone's copy and the iTunes copy, you can go back to the Store and download it at no extra charge.

If you delete a webclip, you can reinstate it by restoring from an iTunes backup (see p.84), though it is generally easier just to go back to the website in question and make a fresh one (see p.78).

You can't delete built-in apps, but you can move them out of the way onto a separate screen or into a separate folder.

- **Deleting apps and webclips from the iPhone** Hold down any Home Screen icon until they all start jiggling and then tap the small red ⊗ of the app you want to delete. When you have finished, hit the iPhone's Home button. This will not remove the app from iTunes, so you can always sync it back to your iPhone again later.

- **Deleting apps from a Mac or PC** Highlight Applications in the iTunes sidebar and then right-click the unwanted app and choose Delete from the menu that appears. If the app was also present on your iPhone, next time you connect expect a prompt to either remove it from there too, or copy it back into iTunes.

> **Tip: When asked to rate the app you are deleting, take the opportunity to give it a score; these ratings are then aggregated back into the App Store for the benefit of other**

07

Maintenance

Troubleshooting and battery tips

The iPhone is a tiny computer and, just like its full-sized cousins, it will occasionally crash or become unresponsive. Far less common, and much more serious, is hardware failure, which will require you to send the phone away for servicing. This chapter gives advice for both situations, along with tips for maximizing battery life.

Crashes and software problems

Every now and again you should expect your iPhone to crash or generally behave in strange ways. This will usually be a problem with a specific application and the iPhone will simply throw you out of the app and back to the Home Screen. From there, simply tap your way back to where you were and start again. If the screen completely freezes, however, try the following steps:

- **Force-quit the current application** Hold down the Sleep/Wake button until the red slider appears, then immediately let go of that button and hold the Home button for five or six seconds.

• **Reboot** As with any other computer, turning an iPhone off and back on often solves software glitches. To turn the phone off, press and hold the Sleep/Wake button for a couple of seconds and then slide the red switch to confirm. Count to five, and then press and hold the Sleep/Wake button again to reboot.

• **Reset** If that doesn't do the trick, or you can't get your phone to turn off, try resetting your phone. This won't harm any music or data on the device. Press and hold the Sleep/Wake button and the Home button at the same time for around ten seconds. The phone may first display the regular shutdown screen and red confirm switch; ignore it, and keep holding the buttons, only letting go when the Apple logo appears.

• **Reset All Settings** Still no joy? Resetting your iPhone's preferences could possibly help. All your current settings will be lost, but no data or media is deleted. Tap Settings > General > Reset > Reset All Settings.

• **Erase all content** If that doesn't work, you could try deleting all the media and data, too, by tapping Settings > General > Reset > Erase All Content and Settings. Then connect the iPhone to your computer and restore your previously backed-up settings (see box on p.84) and copy all your media back onto the phone.

Tip: If you are having problems with Wi-Fi connections, try tapping the Reset Network Settings button on the Reset screen described above.

• **Restore** This will restore the iPhone's software either to the original factory settings or to the settings recorded in the most recent automatic backup (see box overleaf). In both cases all data, settings and media are deleted from the phone. Connect the iPhone to your computer and, within the Summary tab of the phone's options pane, click Restore and follow the prompts.

• **Firmware update** The iPhone's internal software, or firmware, should update itself automatically (see box below). But you can check for new versions at any time by opening iTunes, connecting your phone and, on the summary screen, clicking "Check for Update". If a new version is available, install it.

iPhone firmware

Just as it's always a good idea to run the latest version of iTunes on your Mac or PC, it's also worth running the latest version of the iPhone's internal software – also known as firmware. The phone's firmware will be automatically updated from time to time when your computer is online and connected to the phone. This doesn't affect your settings, media or data, but it fixes bugs and will make the iPhone run more smoothly.

Backing up

iTunes automatically creates a backup of key data on your iPhone whenever you connect to your computer. This can be useful if, for example, you lose your phone and have to replace it with a new one. The backup includes mail settings, SMS messages, notes, call history and Favorites list, sound settings and other preferences. It doesn't include music, video and picture files (since these are already stored in your iTunes Library), photos (which should also be safely residing on your computer) or contacts (which, hopefully, you will have copied across to Address Book or Outlook).

To view, and if necessary delete, an automatic iPhone backup, open iTunes Preferences and click iPhone in the strip along the top. To restore to a backup, simply connect an iPhone and click Restore in the iPhone Summary pane within iTunes.

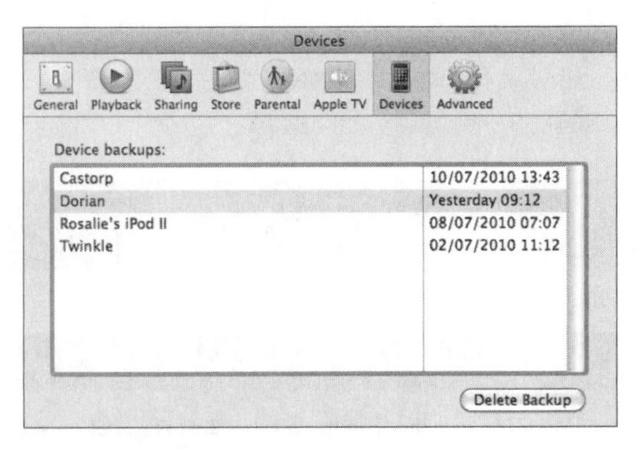

Of course, the backup – and indeed all your media files – is only as safe as your Mac or PC. Computers can die, get destroyed or be stolen, so get into the habit of backing up to either an external hard drive or writable DVDs. For more on this subject, see *The Rough Guide to Macs & OS X*.

The iPhone battery

There has been a fair amount of controversy, and no shortage of misinformation, about the iPhone's non-user-replaceable battery (see p.18). Like all lithium-ion batteries, the one inside the iPhone lasts for a certain amount of time before starting to lose its ability to hold a full charge. According to Apple, this reduction in capacity will happen after around 400 "charge cycles".

Despite various newspaper reports to the contrary, a charge cycle counts as one full running down and recharging of the battery. So if you use, say, 20 percent of the battery each day and then top it back up to full, the total effect will be one charge cycle every five days; in this situation the battery would, theoretically, only start to lose its ability to hold a full charge after quite a few years: 5 days x 400 charge cycles = 2000 days.

If, on the other hand, you drain your battery twice a day by watching movie files while commuting to and from work, you might see your battery deteriorate after just nine months or so. In this case, you'd get it replaced for free, as the iPhone would still be within warranty.

If the battery won't charge

If your iPhone won't charge up via your computer, it could be that the USB port you're connected to doesn't supply enough power or that your Mac or PC is going into standby mode during the charge. To make sure the phone is OK, try charging via the supplied power adapter. If this doesn't work either, it could be the cable; if you have an iPod cable lying around, try that instead, or borrow one from a friend. If you still can't get it to charge, send the phone for servicing (see p.91).

Tips for maximizing battery life

Following are various techniques for minimizing the demands on your iPhone battery. Each one will help ensure that each charge lasts for as long as possible *and* that your battery's overall lifespan is maximized.

- **Keep it cool** Avoid leaving your iPhone in direct sunlight or anywhere hot. Apple state that the device works best at 0–35°C (32–95°F). As a general rule, try to keep it at room temperature.

- **Keep it updated** One of the things that firmware updates can help with is battery efficiency. So be sure to accept any iPhone software updates on offer through iTunes.

- **Drain it** Although it's generally best to leave the device plugged in whenever possible, as with all lithium-ion batteries, it's a good idea to run your iPhone completely flat at least once a month and then fully charge it again.

- **Dim it** Screen brightness makes a big difference to battery life, so if you think you could live with less of it, turn down the slider within Settings > Brightness. Experiment with and without the Auto-Brightness option, which adjusts screen brightness according to ambient light levels.

- **Lock it** Press the Sleep/Wake button when you've finished a call to avoid wasting energy by accidentally tapping the screen in your pocket before the phone locks itself.

- **Turn off Wi-Fi & Bluetooth** These are both power-hungry features which are easily turned off when not in use. Use the switches under Settings > Wi-Fi, and Settings > General > Bluetooth.

• **Junk the EQ** Set your iPhone to use Flat EQ settings (under Settings > iPod). This will knock out imported iTunes EQ settings, which can increase battery demands.

• **Stay lo-fi** High-bitrate music formats such as Apple Lossless may improve the sound quality (see p.139), but they also increase the power required for playback.

• **Slow down** Switch to the slower but more energy-economic EDGE network when battery life is critical (see p.68).

When the battery dies

When your battery no longer holds enough charge to fulfil its function, you'll need to replace it. The official solution is to send the phone to Apple, who, if your phone is no longer within its

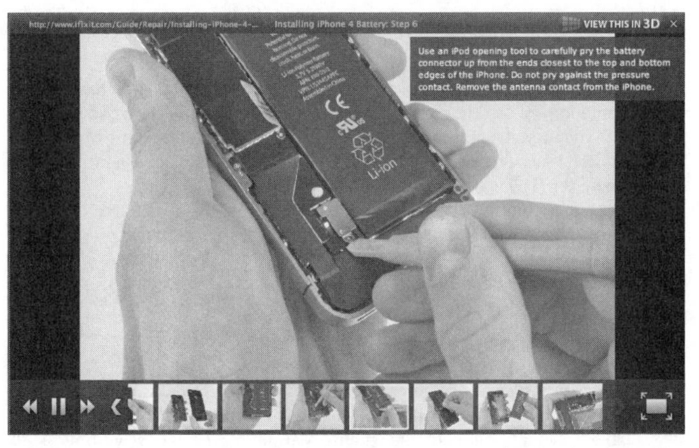

The iPhone 4 battery, pictured as part of iFixit's guide to disassembling the device. For the full guide, see: ifixit.com/Guide/iPhone.

warranty period, will charge you around $75/£55 plus shipping. You'll get the phone back after three working days, but if you can't wait that long, you could hire a replacement for $29/£20.

The unofficial solution is to try to find a less expensive battery from a third-party service, along with a fitting service or DIY instructions. At the time of writing, this isn't an option, but it seems likely that the companies which already offer iPod battery replacements will soon expand their range. These include:

iPod Battery ipodbattery.com
iPod Juice ipodjuice.com

Warranty, AppleCare and insurance

The iPhone comes with a one-year warranty that covers everything you'd expect (hardware failure and so on) and nothing that you wouldn't (accidents, loss, theft and unauthorized service). The only departure from the norm is that while most Apple mobile products come with a warranty that enables you to take the faulty device to any Apple dealer around the world, the iPhone warranty currently only covers the country in which you purchased it.

In addition, you can choose to extend your warranty by a further two years through the AppleCare scheme. The price isn't confirmed at the time of writing, though is expected to be around $60. Assuming AppleCare will include iPhone battery cover, that isn't a bad deal for anyone who expects to regularly use their phone for power-hungry activities such as video watching.

AppleCare apple.com/support/products

As for insurance against accidental damage and theft, AT&T surprised many commentators by not offering an iPhone insurance option at the time of launch – though most participating carriers in most countries do offer insurance, albeit for a fairly high fee. Alternatively, you could investigate what options are available via your home-contents scheme. Many insurers offer away-from-home coverage for high-value items – though this, too, can be expensive.

iPhone repairs

If the advice in this chapter hasn't worked, try going online and searching for help in the sites and forums listed on p.246. If that doesn't clear things up, it could be that you'll need to have your phone repaired by Apple. To do this, you could take it to an Apple retail store (see p.34), though be warned that in some cases you may have to make an appointment.

Alternatively, visit the following website and fill out a service request form. Apple will send you an addressed box in which you can return your phone to them. It will arrive back by post.

Apple Self-Solve selfsolve.apple.com

Either way, you're advised to remove your SIM card before sending it off (see p.91) and you can choose to hire a replacement iPhone for the period of repair ($29 in the US, £20 in the UK). But make sure you return it on time, or you'll be charged a $50 or £25 fee, or the full price of the phone, depending on the length of the delay.

Diagnostic codes

Many phones respond to diagnostic codes – special numbers that, when dialled, reveal information about your account, network or handset on the screen. Assembled in part from a post at The Unofficial Apple Weblog (tuaw.com, essential reading for all Mac, iPod and iPhone users), the following list includes many such codes for the iPhone. Some are specific to AT&T or the iPhone, while others will work on any phone.

Note that these are American codes. European iPhone users may find that only a few of them work on their handsets.

- ***#06#** Displays the IMEI – the handset's unique identification code.

- ***225# Call** Displays current monthly balance.

- ***646# Call** Displays remaining monthly minutes.

- ***777# Call** Displays account balance for a prepaid iPhone.

- ***#61# Cal** Displays the total numbers of calls forwarded to voicemail when the phone was unanswered.

- ***#62# Call** Displays the total numbers of calls forwarded to voicemail when the phone had no service.

- ***#67# Call** Displays the total numbers of calls forwarded to voicemail when the phone was engaged.

- ***#21# Call** Displays various Call Forwarding settings.

- ***#30# Call** Displays whether or not your phone is set to display the numbers of incoming callers.

- ***#43# Call** Displays whether call waiting is enabled.

- ***#33# Call** Displays whether call barring is enabled.

- ***3001#12345#* Call** Activates Field Test mode, which reveals loads of hidden iPhone and network settings and data. Don't mess with these unless you know exactly what you're doing.

Phone

08

Contacts

Importing, syncing & managing

Just like every other mobile phone, the iPhone allows you to store a list of names and phone numbers. Unlike some other phones, it also lets you store all kinds of other information for a contact – from their email and fax to home address, birthday and job title. You can fill in as much or as little as you like. Even better, the iPhone makes it easy to synchronize your contacts info with the address book on your computer. This is not only convenient, but means that your precious names and numbers can be safely backed up.

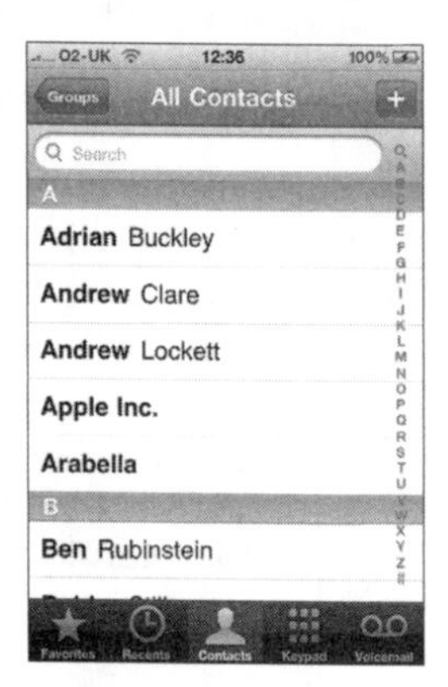

If you've been living on the moon for the last decade, you might wish to start your contacts list from scratch, in which case turn to p.105. Probably, however, you'll want to import the contacts from your old phone or computer…

Importing contacts from a phone

At the time of writing, the iPhone can't communicate with other mobile phones via Bluetooth and it won't read most SIM cards. So your best bet for importing contacts from an old phone is to copy them across from the old phone to your computer, into one of the address book applications that the iPhone can communicate with. This way you'll also be able to tidy up and consolidate your contacts info on your computer before moving everything over to the iPhone.

Of course, if you only have a few numbers on your old phone, or the following techniques seem like too much hassle, you could simply enter your numbers into your computer manually (see p.100) or iPhone (see p.105).

Moving contacts from an old phone to a Mac

With Bluetooth

First you need to pair your phone and Mac. To do this, make sure the phone is "discoverable", an option usually found under Connectivity or Bluetooth. Next, on your Mac, click…

> System Preferences > Bluetooth > Devices > Set Up New Device

… and follow the prompts. Once that's all done, open iSync from your Mac's Applications folder and choose Add Device from the Devices menu. With any luck, your phone will be recognized and appear on the toolbar. Choose what you want to copy across

Tip: Most Macs and many PCs shipped in the last few years have Bluetooth built in; for those which don't, this feature can also be added with an inexpensive USB adapter.

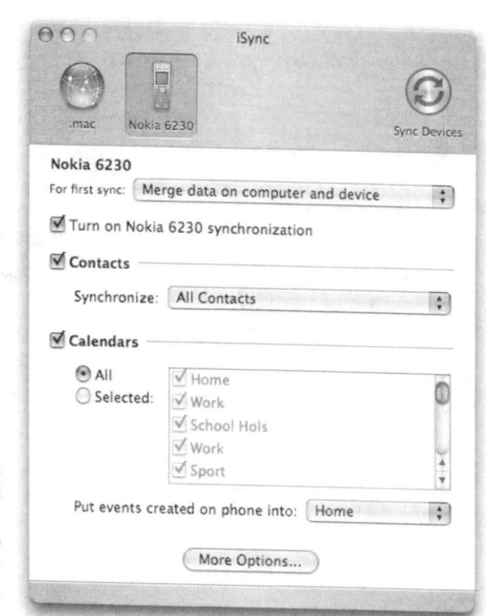

Apple's iSync application, included on all recent Macs, makes it simple to import numbers from mobile phones.

(contacts, calendars or both), hit the Sync Devices button and the data should be imported from the phone into your Mac's Address Book, ready for transfer to your iPhone. Turn to p.100 for more on using Address Book.

Don't have Bluetooth?

If your phone or Mac lack Bluetooth, you could buy the appropriate cable to connect your phone directly to your Mac (ask in any phone store) and use iSync as described above.

Alternatively, you could try a SIM reader to extract vCard or text files from the phone. These files can then be imported into the address book on your computer, ready to be synced with the iPhone.

SIM readers

A SIM reader is an inexpensive device that will feed numbers stored on a mobile phone SIM card into a Mac or PC via a USB socket.

Before using a SIM reader, you'll need to make sure that the numbers on the phone are stored on the SIM rather than in the phone's memory. Consult the manual to find out how this is done (you can probably download the manual if you no longer have it). Then remove the SIM, insert it into the reader and connect it to your computer. You'll probably be left with vCard or text files, which should be easy to import into Address Book, Outlook or Outlook Express (look for the Import option in the File menu).

If you're unlucky, and you can't get the files to import, try opening them with Text Edit (Mac) or Notepad (PC). You should then at least be able to see the data from your phone and, if needs be, copy and paste it into new contact cards.

Moving contacts from an old phone to a PC

The iPhone can pull contacts from either Outlook Express (ie the Windows XP Address Book) or Outlook on a PC. Unfortunately, getting numbers from your phone into one of these applications can be tricky, because Windows doesn't ship with a simple tool for moving data from phone to PC.

If your phone came with a CD, it may contain software that will let you export the numbers to your phone as vCards or text files, either via Bluetooth or a cable. In most cases, these files can be dropped into Outlook or the Windows XP Address Book (choose Import in the File menu).

Alternatively, you could buy a downloadable number-transfer tool such as the following. These cost $20–40 but will help you copy across photos and movies, as well as contact info:

DataPilot datapilot.com
Mobile Master mobile-master.com
WinFonie Mobile 2 bertels-hirsch.de/en/winfonie_mobile_2

Finally, you could try a SIM reader (see box, p. 98) or ask for help in a phone store. For a small fee, some stores will pull the numbers off your old phone and provide them on a CD or flash drive, ready to be imported into your contact application of choice. Similar services are available online from sites such as:

CellularDr cellulardr.com (US only)

From old phone to the Yahoo! Address Book

iTunes can sync contact info between your Yahoo! Mail address book and your iPhone. So one final option for getting the numbers off your old phone is to upload them directly to Yahoo! If you currently have a smartphone or handheld device that runs Palm OS, this can be done by using Yahoo!'s Intellisync feature – look for the "Sync" button on your Yahoo! Mail homepage to see how it can be set up. Regular mobile phone users can take a look at what's on offer from the Yahoo! Mobile service (mobile.yahoo.com), though not all handsets are compatible.

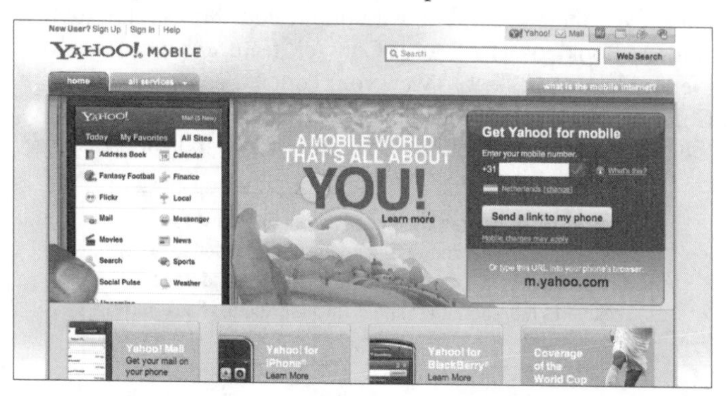

Editing contacts on a computer

However you get the numbers from your old phone to your computer, it's likely that the result will need some editing – duplicates culled, email addresses and other information added, and so on. All this can be done on the iPhone (as we'll see), but it's easier on a computer.

It's beyond the scope of this book to walk you through contact management in Address Book, Outlook and Outlook Express, but here are some general pointers:

- **Default formats** Look within your Preferences or Options to check that the default address and phone number format is correct for the country where you live.

- **Groups** You can add contacts in your address book to "groups" or "distribution lists". You might have all your work colleagues in one group, say, and all your friends and family in another. When you sync your iPhone you can choose to copy across either all contacts or just selected groups.

- **Duplicate entries** After you import the contacts from an old phone, you may well end up with duplicated entries. The Apple Address Book boasts a "Merge Contact" feature that can resolve the problem: click ⌘+1 to view your contacts as Cards and Columns; next, search for the duplicated contact's name, select the two entries when holding down the ⌘ key and choose Merge

Tip: **On a Mac, you can choose to have all the websites associated with your contacts in Address Book added to your bookmarks in Safari. In Safari Preferences, under the Bookmarks tab, check the various "Include Address Book" boxes.**

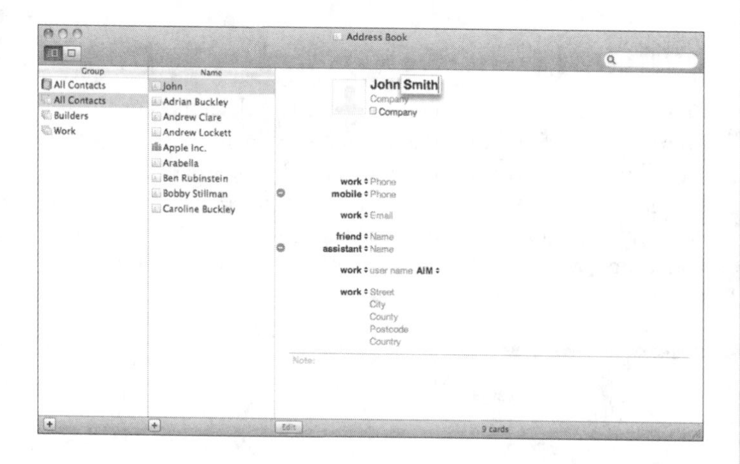

Selected Cards from the Cards menu. Outlook, meanwhile, will automatically alert you when duplicate entries appear and ask you how you want to resolve the conflict. To check whether this feature is enabled, open Options from the Tools menu.

• **Pictures** Though time-consuming and not particularly useful, adding images to contacts can be a fun way to personalize your Address Book on your computer and, in turn, your iPhone. It can be especially useful if you are the kind of person who finds it hard to put names to faces. You can, of course, take snaps out and about with the iPhone itself and associate those images with your contacts there and then (see p.204), but pictures can also be added to your computer's address book. On a Mac, simply drag a photo onto an entry. Of course, you don't even have to use mug shots: if, like both Apple and Microsoft, you think an individual might be best represented by, say, a tennis ball or a butterfly, then so be it.

Syncing contacts with the iPhone

Open iTunes, connect your iPhone and select its icon in the Source panel. You'll find the Contacts options under the "Info" tab. Check the relevant sync option and choose whether you want to import all contacts or just particular groups. Once that's done, each time you connect your iPhone, any new contact information added on the computer will be copied across to the phone, and vice versa.

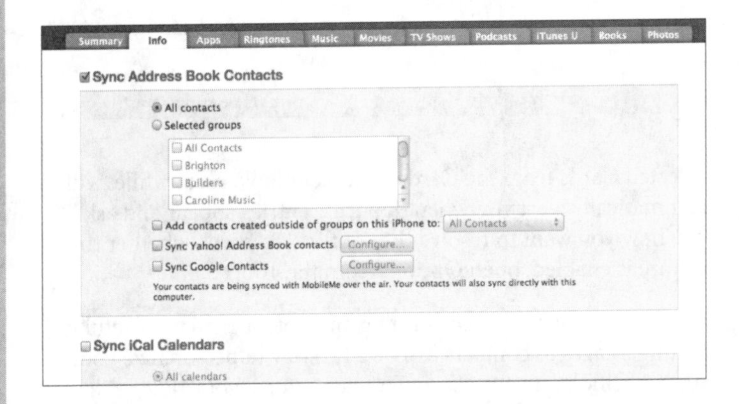

Syncing contacts with Entourage

Apple describe Entourage – the Mac version of Outlook – as being compatible with the iPhone in terms of syncing contacts and calendars. But this isn't possible directly. In order to enable syncing between the iPhone and Entourage, you first have to enable syncing between Entourage and the Apple Address Book. To do this, open Entourage and choose Preferences from the Entourage menu. Under Sync Services, check Contacts and Events. If you can't find these options, you may need to update your software. Click Check for Updates in the Entourage Help menu.

In addition, if you use Yahoo! Mail or Gmail you may want to opt to import the email addresses and other contact information stored in your Yahoo! or Google address book. If you select Yahoo! or Google *and* another contacts source, you'll find that these two sources become synchronized with each other, so you'll end up with the same amalgamated contacts database on your phone, your Yahoo! or Gmail account and your computer.

Plaxo: syncing multiple contacts

As we've seen, iTunes can sync with contacts in Address Book, Outlook and Outlook Express, and Yahoo! Mail. If you have contact info in Gmail, AOL, Hotmail or Thunderbird, you'll be interested in Plaxo, a free online contact and calendar management tool co-created by one of the names behind Napster, Sean Parker. Plaxo can merge and update address books from all the services just mentioned, in addition to those which iTunes can handle. So you could use it to get all your contacts onto your PC or Mac, and from there onto your iPhone.

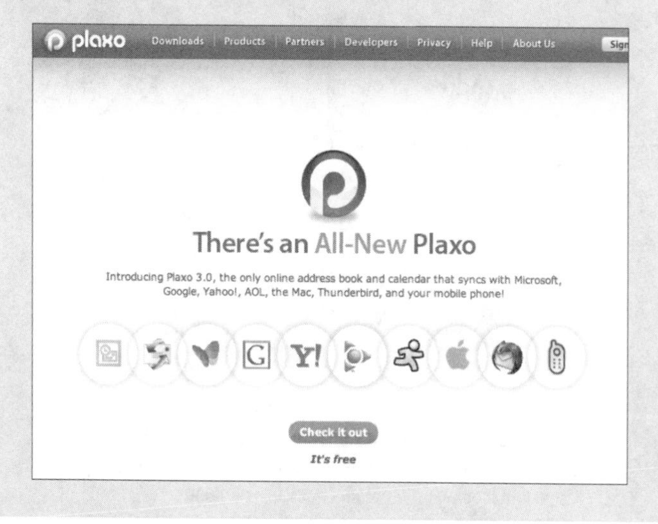

Contacts on the iPhone

You'll find the iPhone's main Contacts list as one of the five buttons when you tap Phone. To browse the list, either flick up or down with your finger, or drag your finger over the alphabetic list on the right to quickly navigate to a specific letter – useful if you have an extensive list of contacts. Alternatively, tap the search box at the top of the screen and type in the first few characters from the name you're looking for.

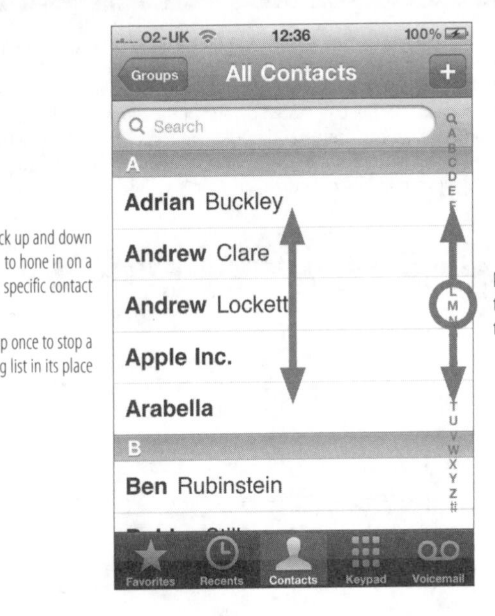

Flick up and down to hone in on a specific contact

Tap once to stop a moving list in its place

Drag up and down the alphabet to speed to a specific letter

Tip: By default the iPhone alphabetizes your contacts by surname. You can change this by clicking Settings on the main screen and looking under Mail, Contacts, Calendar.

When you find the contact you want, tap once to view all their details, and then tap a phone number to start a call or an email address to launch a new mail message.

Adding contacts

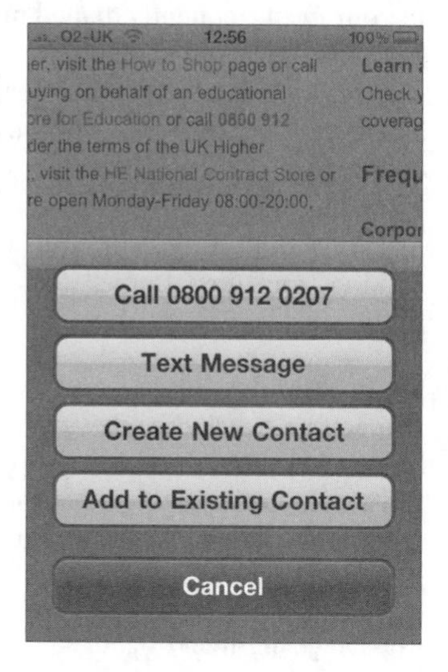

There are various ways to add new contacts on the iPhone.

- **In Contacts** When viewing your contacts list, click the **+** at the top right.

- **From the keypad** You can enter a number via the keypad (found under Phone) and then tap the head-and-shoulders graphic to the left of the call button. From there you can choose to create a new contact or add the number to an existing contact.

- **From the recent calls list** When viewing recent calls (under Phone), you can tap the ⓘ icon next to an unrecognized number and choose to create a new contact or add to an existing contact.

- **From an email or webpage** If someone emails you a number, or you stumble across someone's number online, tap and hold it until the options to either Create New Contact or Add To Existing Contact appear.

> **Tip:** **You can also add contacts to your list by photographing business cards using a downloadable app such as BC Reader.**

Favorites

The Favorites list, found from the Phone screen, provides quick access to your most frequently dialled numbers. Instead of full contacts, it stores a specific number for each name. This way you can dial with just a couple of clicks – much quicker than browsing through a long contacts list, selecting a person and then picking from their home, work and mobile numbers.

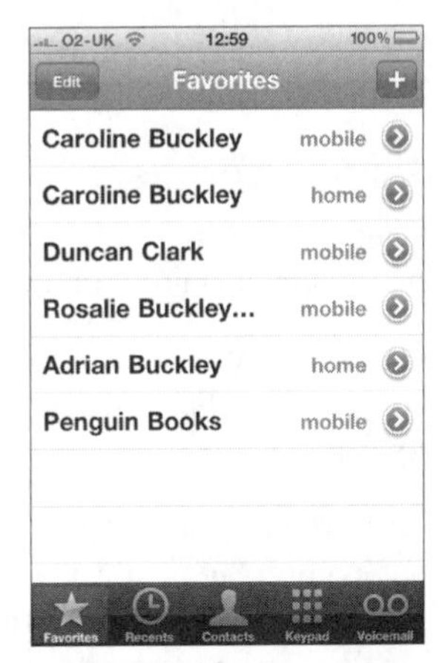

• **To add Favorites**, browse your contacts and click the relevant names, in each case choosing Add to Favorites and picking a number. Alternatively, from Favorites, hit the **+** button to browse for names.

• **To change the order** of your Favorites, tap the Edit button and drag contacts by their right-hand edge (pictured).

• **To remove a Favorite** click Edit, tap the red ⊖ icon next to the relevant name, and then hit Remove.

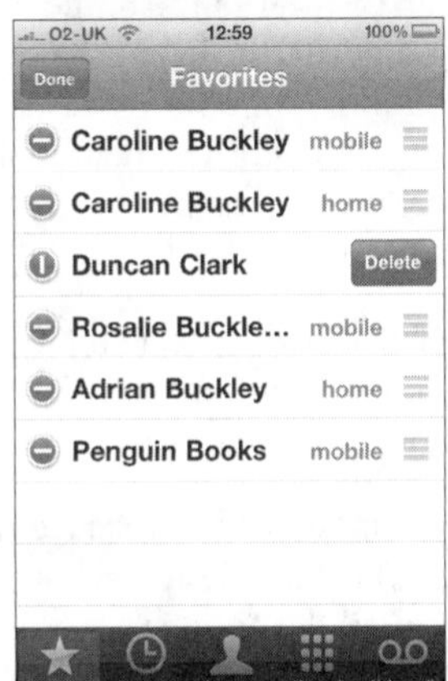

Tip: **When a number is added as a Favorite, a small star appears next to the relevant number on the contact's page.**

Editing contacts

To edit a contact – change their name or number, assign a specific ringtone, add an email address, or whatever – simply click the relevant name in the main contacts list and hit Edit. Then…

• Use the Add options, signified by a green ⊕, to add a new number, address or other attribute.

• If you can't see the relevant attribute, click Add Field. You'll then be offered everything from birthday to a space for notes. This info may then be useful back on your computer. For instance, if you add birthdays, these can sync with your calendar application, which you could set up to automatically send you a reminder email a day or two before each friend or relative's birthday.

• To delete an item, tap its red ⊖ icon. To delete the contact from your Contacts list entirely, scroll down to the bottom of the entry and tap the "Delete Contact" button.

> **Tip:** **Syncing with Yahoo! won't delete any contact entry in the Yahoo! Address Book that contains an ID for Yahoo! Messenger, even if you delete that contact on the iPhone or your computer. To delete such a contact, log in to your Yahoo! account and remove it directly.**

• To assign a picture to a contact, click Add Photo (top left) and choose either to take a photo with the iPhone's camera or choose an existing photo from the albums or camera roll already stored on the phone. Once you've selected an image, you can pinch and drag to frame the snap just the way you want it. When you're done, click Set Photo.

> **Tip: Adding pauses to phone numbers. If you're adding a phone number that needs a pause at a certain stage – such as before a passcode or extension number – tap the # key, then tap Pause.**

Contacts can...

Aside from making calls, contacts on the iPhone have several uses, including:

• Tap a contact's email address to instantly be presented with a blank email message addressed to that contact from your default account. For more on email, see p.187.

• Tap a contact's web address listing to have the iPhone switch to Safari and take you straight to that webpage. For more on browsing the web with an iPhone, see p.175.

• Tap a contact's address and the iPhone will show you the location in Maps. For more on Maps, see p.209.

• To share a contact's details in the form of a so-called vCard, tap the Share Contact button at the bottom of that contact's details screen and then choose to either send the file via email or MMS.

09

Calls

From conference calling to video chat

Though the iPhone is special in many ways, it largely sticks to the familiar when it comes to making and receiving calls. This chapter whizzes you through the basics, offering some useful tips and tricks along the way.

Making calls

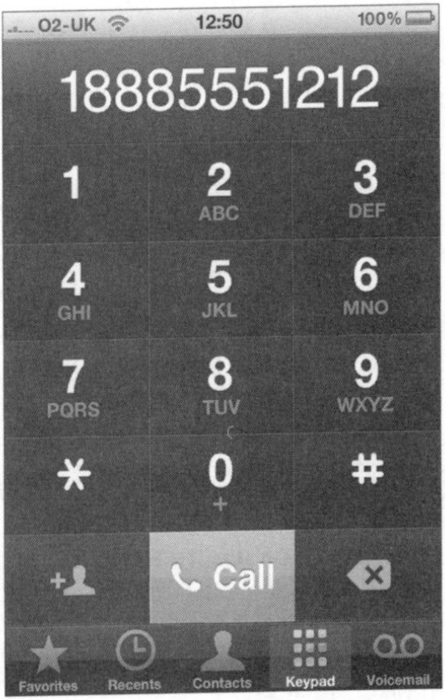

There are various different ways of making a call using the iPhone. Simply tap the Phone button, and then…

• Tap Contacts, pick a name and then choose a number (home, mobile, work, etc).

• Tap Keypad and type a number (hit ⌫ if you make a mistake). Then tap Call. You can even paste in a number by tapping the space just above the keypad and then tapping Paste.

• To bring up and redial the last number you entered manually, tap Keypad, then Call.

• Tap Favorites and hit a name. Because each Favorite is a specific number, this saves you a couple of taps.

• To call someone you recently phoned, or who recently phoned you, click Recents and hit the relevant name or number.

Tip: **Even before your iPhone is activated or if it doesn't have a SIM card inserted, you can still call the emergency services. Tap Keypad and dial, for example, 911 in the US or 999 in the UK, then tap Call.**

When you're on a call...

When the iPhone is close to your ear, a proximity sensor disables the screen, saving you power. Move the phone away from your ear while on a call and the screen displays various call options:

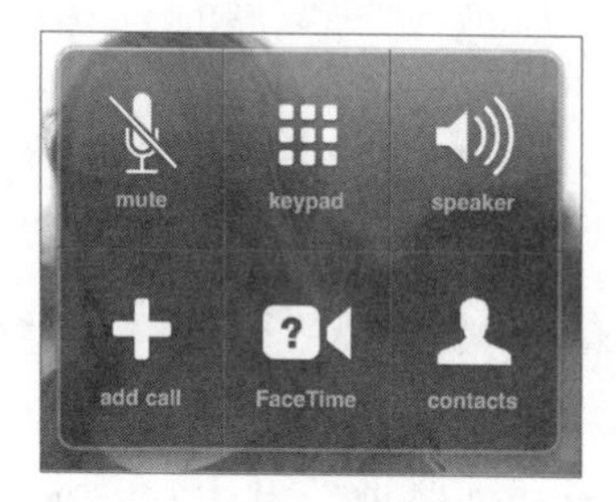

- **Mute** When tapped, the caller can't hear you. Tap again to return to normal.

- **Add Call** For multiple and conference calls (see overleaf).

- **Keypad** Brings up the numeric keypad – essential when using automated phone systems.

- **FaceTime (iPhone 4 only)** Tap to initiate a video call with another iPhone 4 user – see p.120.

- **Hold (when FaceTime isn't available)** Tap once and neither you nor the caller can hear each other. Tap again to return to normal.

- **Speaker** Toggles speaker-phone on and off. The iPhone's speaker is located on the base of the unit. As with other phones, the sound isn't amazing and may distort, in which case reduce the volume level using the buttons on the side of the iPhone.

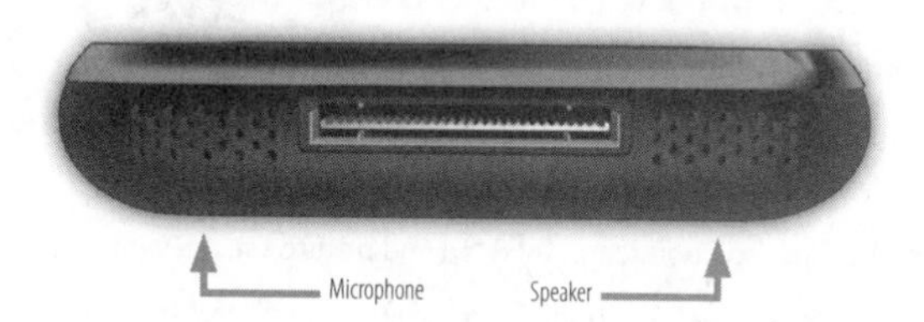

- **Contacts** Takes you to your main Contacts list.

Multiple calls and conference calling

The iPhone has two lines available for calls. When you're talking to one person, you can click Add Call and dial someone else; the existing caller will be put on hold, and you can then use the Swap option to switch between the two lines. Likewise, if someone calls you while you're on the phone, you'll be offered a Hold Call + Answer option.

Even better, the iPhone allows you to make conference calls with up to five people simultaneously. With a conference call, you don't need to switch between callers – everyone can be heard by everyone else. It works like this: one of the iPhone's two lines hosts the conference and the other is free to call people who can then be merged onto the conference line. And so on.

So, to get things started, make or receive a call in the usual way. Next, tap Add Call and dial someone else. The first call is put on hold while you do this so you can give the second person you've called warning that you want to add them to a conference – or, indeed, chat privately. Then, hit Merge Calls to create the three-way conversation. Follow the same procedure to add further callers to the party.

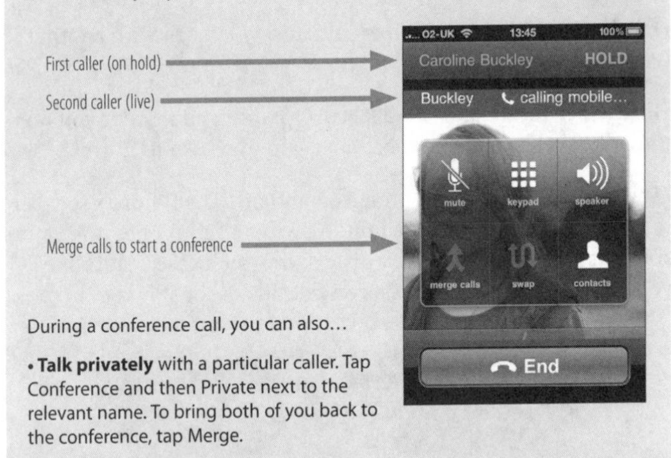

During a conference call, you can also...

- **Talk privately** with a particular caller. Tap Conference and then Private next to the relevant name. To bring both of you back to the conference, tap Merge.
- **Ditch a caller** Tap Conference, then tap End next to the relevant name.
- **Add an incoming call** Tap Hold Call + Answer and then Merge Calls.

And more...

While on a call, you can press the Home button to access any other applications without dropping your call – useful if you need to check a date in your calendar or an address in an email. Note, however, that you can only access the Internet during a call when you're connected to a Wi-Fi network (see p.65). So don't expect to use Safari or Maps while walking down the street in the middle of a conversation.

When you have finished with whatever it was you were doing, press the Home button again and then tap the green strip at the top of the screen to return to the Call Options screen.

Receiving calls

When someone calls you, the iPhone will either ring or vibrate (for setting up ringtones, etc, see p.52) and display the caller's information on the screen, including their photograph, if you have one set up in Contacts (see p.204). Next, do one of the following:

To answer a call

To talk when a call comes in, tap Answer or, if the iPhone is locked, drag the slider. If, when the call comes in, you have audio or video playing, it will fade out and pause. If you're using the supplied headset, click the headset's mic button.

To decline a call

If you don't want to talk, declining a call will send it straight to voicemail. This can be achieved either by tapping Decline on the screen, or:

• Pressing the Sleep/Wake button on the top of the iPhone twice in quick succession.

• If you are using the iPhone headset, press and hold the mic button for a couple of seconds, then let go. You will then hear two low beeps confirming that the call has been declined.

To silence a call

When a call comes in, you can quickly stop the iPhone ringing or vibrating without answering or declining the call – useful if, for instance, you get a personal call in the office and you want to step outside before answering it.

To do this, simply press the Sleep/Wake button or either of the volume buttons.

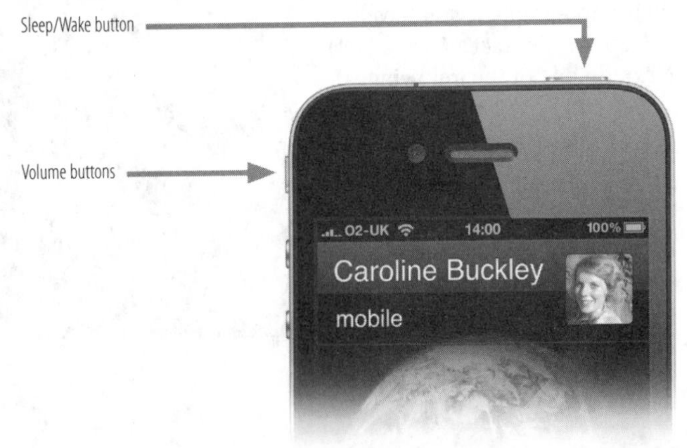

To answer a second call

If you have Call Waiting switched on (within Settings > Phone), you can receive a second call while already on the phone. The iPhone will chirp in your ear, show the new caller information and offer you three options:

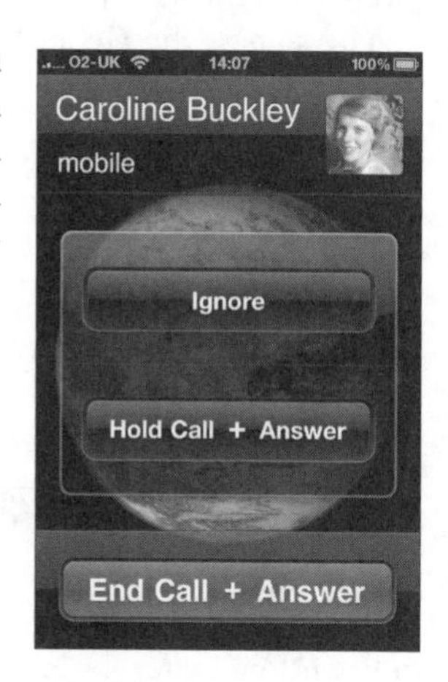

• **Ignore** Sends the new caller to voicemail.

• **End Call + Answer** Ends the call you were on and answers the new one.

• **Hold Call + Answer** Puts the first call on hold and answers the second. From there you can either switch between the two conversations using the Swap button or hit Merge to combine the calls into a three-way conference.

You can turn Call Waiting on and off within Settings, under the Phone sub-menu.

Recent and missed calls

Like other mobile phones, the iPhone keeps a list of recent incoming and outgoing calls. This can be used for reference – for example, showing you the time when someone phoned – or as a means of storing numbers or making calls. To access the list, tap Phone > Recents. Note that the Phone icon displays a small red circle containing the number of missed calls and unheard voicemails you have, and Recents boasts a similar circle only listing missed calls.

In the list, missed calls appear in red and can be viewed in isolation by tapping the Missed button. When a caller has attempted to reach you more than once, the number of missed calls is displayed in brackets.

Tapping ⊙ to the right of any entry will display more information about the call, such as whether it was incoming or outgoing. When a caller is already in your Contacts list, all their information is displayed, with the number that relates to that specific call highlighted in blue.

Visual Voicemail

One of the most innovative features ushered in by the original iPhone was Visual Voicemail. The idea is that you no longer have to listen to all your voicemail messages in turn to get to the one you want. Instead, your voicemails are presented in a list – just like emails – and you can choose the ones you want to listen to in any order. You can even rewind and fast-forward. Use the system for just a few weeks and you will hardly believe it was ever done the old way.

As mentioned, the Phone button on the Home Screen displays a red circle containing the total number of missed calls and unheard voicemails. Tapping it reveals the Voicemail button, which also displays a red numbered icon, but this time just for unheard voicemails.

Voicemail setup

Tapping the Voicemail button for the first time takes you to a screen with an option to create a voicemail password, which you can use to access your voicemails from other phones (see p.119); you can change this password at any time by tapping through to Settings > Phone > Change Voicemail Password.

Voicemail overseas

In some foreign countries, you may find that Visual Voicemail won't work. Instead, when you click Voicemail, you'll be offered a single Call Voicemail button which will take you through to your messages the old-fashioned way. As with other phones, you can also call your voicemail by holding down the 1 on the numeric keypad.

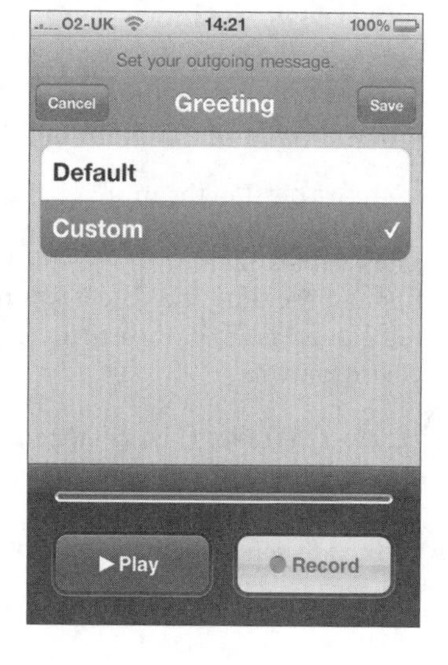

The first time you tap Voicemail, you'll also be prompted to record a greeting, which callers will hear prior to leaving you a message. Tap Greeting, then Custom, then Record. When you're done, you can play back your message and, if you're happy with it, tap Save. Alternatively, if you're feeling shy, you could tap Default instead of Custom and stick with the pre-recorded greeting, which includes your number.

By default, the iPhone will alert you with a sound when you have a new voicemail (except if the Silent switch is on). If you'd rather turn this function off, tap Settings > Sounds and set the New Voicemail switch to Off.

Tip: Messages are saved for thirty days from the time you first listen to them – regardless of whether or not you delete them.

Playback and more

The Voicemail screen lists current voicemail messages, with those you have not yet listened to displaying a blue dot to their left. It is worth knowing that even though a voicemail appears in the list, it hasn't been downloaded to the phone. Thus, you need to have a network signal to listen to voicemails.

Each voicemail lets you…

- **Play/pause/rewind** You can play and pause a message at any time with the ▶ and ❚❚ icons. Even better, you can rewind or skip forward by dragging the scrubber bar. (One of the iPhone's finest moments in terms of innovation.)

- **Return a call** Tap the message and hit Call Back.

- **View details** Tapping the ⊙ button to the right of any message to find out the time and date it was recorded, its duration and the full contacts info of the caller, where known.

- **Contact the caller** Having tapped ⊙, you can tap the caller's number to call them, email address to send an email, or Text Message to send an SMS.

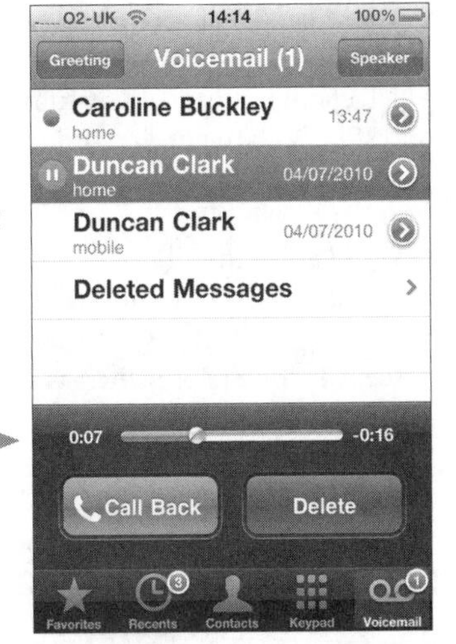

Scrubber bar

• **Add to Contacts** Tap ❯ next to a message and then Create New Contact, Add to Existing Contact or Add to Favorites.

Deleting and undeleting voicemails

One great thing about the iPhone's voicemail system is that deleted messages are saved for thirty days before being permanently erased.

• **To delete a message** Tap a message, then tap Delete.

• **To view deleted messages** Scroll to the bottom of the voicemail list and tap Deleted Messages.

• **To undelete a message** Choose a deleted message and click Undelete.

Picking up voicemail from another phone

You can pick up your iPhone voicemails the old-fashioned way using any phone. At least, American iPhone owners can; it remains to be seen if this will work in the UK or elsewhere.

Simply call your iPhone's number and, assuming it is not answered, you'll be redirected to your voicemail. When you hear your greeting, dial * followed by your voicemail password. Then enter # and follow the voice instructions.

Voicemail over speaker and Bluetooth

To listen to your voicemail messages over the iPhone's built-in speaker, tap the Speaker button in the top-right corner. Or, if your iPhone is connected to a Bluetooth headset or car kit (see p.237), tap Audio and choose Speaker Phone to use the built-in speaker. To switch back to the headset or car kit, tap Audio again, then choose the relevant device.

FaceTime

Easily the most important feature added by the fourth-generation iPhone was FaceTime, a foolproof way of making and receiving video calls. Unimaginable except in science fiction just a decade or so ago, mobile video calls offer a whole new type of communication – the only downside being that it only works when both people on the call have an iPhone 4 or some later model.

Video calling required the addition of an extra camera lens on the front of the iPhone so that the user's face can be captured without spinning the phone around and obscuring the screen. Best of all, because there's a camera lens on the back, too, it's possible at the tap of a button to switch between the two views – one of the caller and one of their view.

Using FaceTime

Assuming FaceTime is switched on (you can check within Settings > Phone), then making a video call is as easy as tapping a button. However, at the time of writing, due to limitations of the 3G networks in many countries, it will usually only work when your phone is connected to the Internet via Wi-Fi (see p.65). Doubtless this will change in time, enabling iPhone users to make video calls while they are out and about.

To initiate a video call, either:

• Tap the FaceTime button on a contact's page in your contacts list.

• Make or receive a phone call in the usual way and then hit the FaceTime button on the call screen (pictured here).

Either way, the person you are calling will immediately receive a FaceTime invitation. If they accept, the video call will begin. During the call, each user can toggle between their iPhone's two cameras using the "switch camera" icon located at the bottom-right of the screen that displays while you are using FaceTime.

Front lens
for showing the caller's face

Rear lens
for showing whatever the caller is looking at

Voice Control

If you have either an iPhone 3GS or iPhone 4 then you can also initiate a call using only your voice. It can be a bit hit and miss, especially when out and about, near noisy traffic, or the like, but is still worth playing around with.

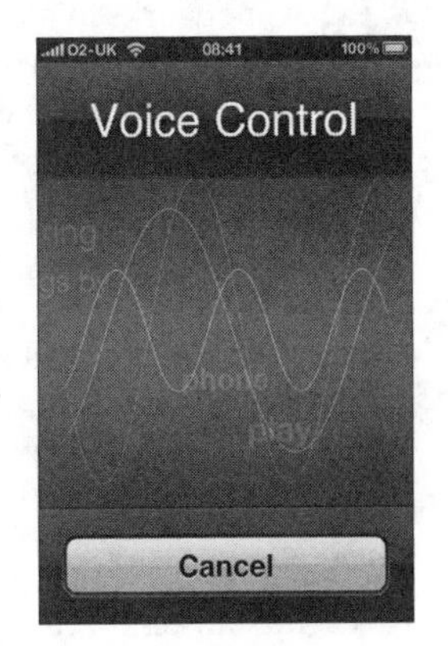

• **To call a contact** Press and hold the Home button for around a second until the Voice Control screen pops up and you hear a noise. Then speak clearly into your phone, saying "call" or "dial", followed by the contact's name. You can also add "at home" or "mobile" if you know the contact has multiple numbers. If you don't do this, your iPhone will next ask you to clarify which number you want.

• **To call a number** Press and hold the Home button until you hear a noise. Then say "call" or "dial", followed by the number, speaking each digit and avoiding words such as "double" and "triple". In the US, however, you can say "eight hundred" for the "800" area code.

This is all best done with the iPhone held away from your ear so that it is easy to see from the screen whether the feature has heard you correctly and is dialling the right number.

Tip: To stop voice control randomly calling people when the phone is locked (and the Home button is accidentally pressed in your pocket or bag) go to Settings > General > Passcode Lock and turn the Voice Dial setting to "off".

Other call features

Call Forwarding

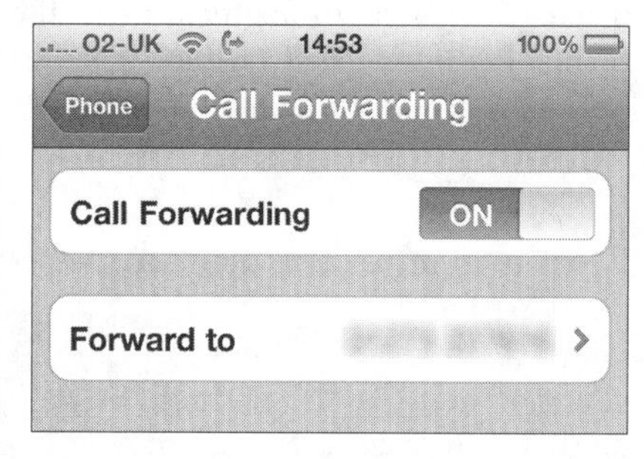

If you'd like to have incoming calls forwarded to another number, click Settings > Phone > Call Forwarding and enter a number. This can be very handy if, for instance, you're going to be outside your network's coverage area but available on a landline. When this feature is active, a special blue icon appears at the top of the iPhone's display to the left of the clock.

Caller ID (outgoing)

If you'd like to call someone without your name or number flashing up on their phone screen, click Settings, then Phone, and switch off Show My Caller ID.

Super Caller ID

Tap a received call in your Recents list and you'll see a new screen with options for calling back, adding to contacts, etc. This page includes address details, when known, for callers already in your contacts list. For unknown callers, if you're in the US, you'll see the area where the call came from. This "Super Caller ID" feature can be very handy when you receive a call with an unfamiliar area code. Unfortunately, the same info doesn't flash up when the phone actually rings.

Calling apps

There are hundreds of apps available in the iTunes App Store that can be used to augment your iPhone calling experience. Many offer pure novelty, while others bring some interesting and genuinely useful additional functionality.

callLog

When used in place of the iPhone's built-in call list interface, this app allows you to add notes and reminders to your calls and also export reports on call activity.

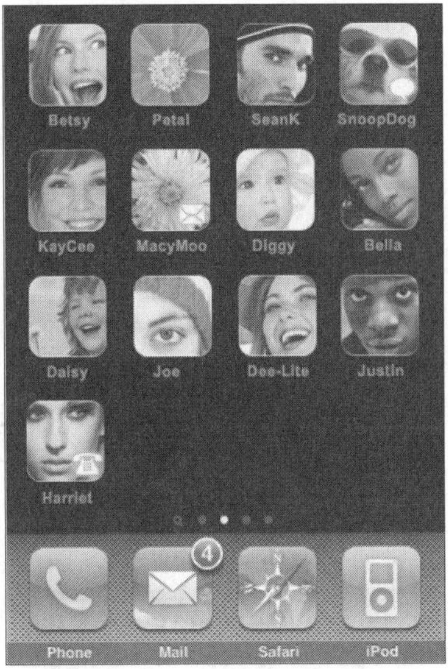

Dial Plate

This is one of many apps that add a retro rotary dialler to your iPhone's armoury.

FaceDialer

A very handy app for adding individual numbers to your Home Screen with a photo icon of the contact (pictured). There are dozens of similar apps available that add icons to the Home Screen for shortcut calling friends and loved ones ... CallHim and CallHer are among the more stylish offerings for speedy spouse dialling.

Calling via the Internet

Anyone used to using Skype or other Internet calling software on their computer will be aware that it's possible to make free or virtually free Internet calls to landlines and mobiles all over the world. Given that the iPhone is essentially a small computer that can connect to the Internet, it's not surprising that it allows you to do just the same. Here are some of the options.

Skype

Easily the most popular Internet calling system, Skype is available on the iPhone via a free application available from the App Store (see p.72). Once it's installed, either log in with your existing Skype account details or set up a new account and you'll immediately have access to all the same features familiar from the computer version of Skype – instant messaging, Skype-to-Skype calls and, if you buy some credit, very inexpensive Skype-to-phone calls.

At the time of writing, Skype-to-Skype calls can be made for free both via Wi-Fi and 3G, though the company has made it clear that calls via 3G will incur a small charge from the start of 2011.

Truphone

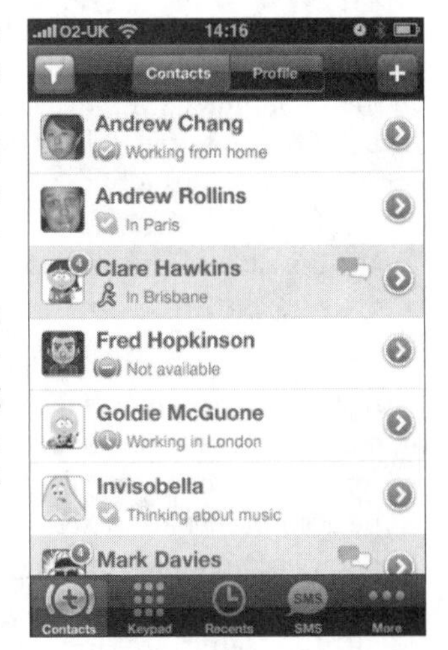

A very good alternative to the Skype app is the free Truphone one. It gives you VoIP call contact with Skype, Truphone and Google Talk users and also a free voicemail service with push notifications.

It works really well over both Wi-Fi and 3G and it's also a multitasking app (see p.77), meaning that it can run in the background, allowing you to receive calls from the service all the time.

Tip: When calling via the Internet, you'll usually need to enter numbers complete with international dialling codes, even if you're not calling overseas. So it's good to get into the habit of using dialling codes when adding or editing contacts.

Calling an email address

Jangl offers a different take on free Internet-based mobile phone calls. Enter an email address of someone you know and Jangl will assign a number to that address. Dial this number and you can leave a voicemail for the person, who will receive it via email – wherever they are in the world. Or, if they're already a member, the call will divert to their nominated phone.

In addition, Jangl offers local-priced calls to around thirty countries, mainly in Europe and the Americas.

Jangl jangl.com

Fring

Similar to Truphone, Fring gives you access to a range of contacts, this time those who are members of Windows Live Messenger, Google Talk, Twitter, Yahoo, AIM and ICQ. The service also allows you to make two-way video calls over its network, with an experience similar to Apple's FaceTime feature (see p.120).

Jajah

Although apps such as Skype and Truphone are the most obvious way to make Internet calls via your iPhone, it's actually one of many options. Sign up for a free Jajah account, for example, and you can call other Jajah users for free – even internationally – and make other long-distance calls at discounted rates (as little as 3¢/1.5p for the US, China and most of Western Europe). You go to the website using Safari and enter your own number and the number you want to call. Press Go and your phone will ring; pick it up and the number you're calling will ring. Neither of you initiated the call, so neither of you will pay long-distance charges.

Jajah iphone.jajah.com

Rebtel

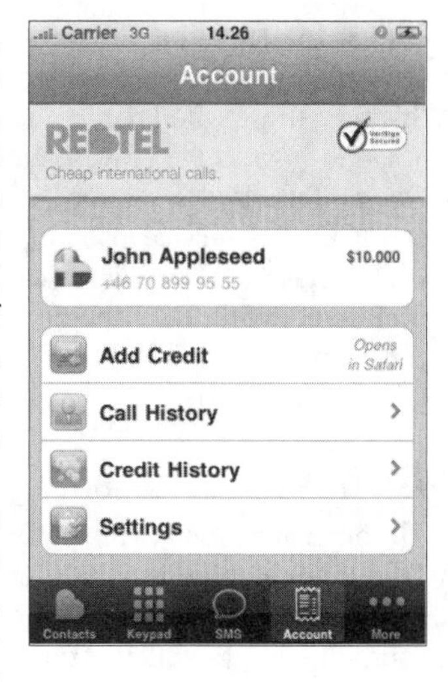

Rebtel works slightly differently. Enter the phone number of a friend, relative or colleague in any of the fifty or so supported countries and Rebtel will give you a local number on which you can call using your free minutes, plus a tiny fee to Rebtel. Best of all, the numbers stick, so you can save one to your phone and use it whenever you want to call the same person.

The fees are as little as 1¢/1p per minute, but you can even avoid paying these if you like, by getting the person you're calling to phone you back on the number that shows up on their screen when you ring. There's also a very nice iPhone app (pictured), which makes the whole process far easier to organize.

Rebtel rebtel.com

Calling from a Mac or PC

One advantage of being able to easily sync all your phone numbers from your phone to your computer is that you can then import them into an Internet telephony program and call the numbers from your computer in order to save your precious mobile minutes. For example, the free desktop version of Skype, which is free to download, lets you import numbers from Address Book or Outlook. Just choose Import Contacts… in the Contacts menu.

Skype skype.com (PC & Mac)

Other programs that let you make cheap calls over the Internet to regular and mobile phones include:

AIM aim.com (PC & Mac)
Windows Live Messenger get.live.com/messenger (PC)
Yahoo! Messenger messenger.yahoo.com (PC & Mac)

To make decent-quality calls, your computer will need an audio headset or USB handset (pictured). It is possible to get by with a regular microphone and speakers (including those built into most laptops), but these tend to lead to annoying feedback and echoes because the sound from the speakers gets picked up by the microphone.

10

SMS

Texting by phone or computer

The iPhone is unprecedented in its incorporation of SMS, short for Single-Molecule Spectroscopy. OK, not really: the SMS application on the iPhone Home Screen refers to Short Message Service – aka text messaging. It works just like on any other phone, except that you get to use a full QWERTY keyboard and see your texts as threaded conversations – just as with chat tools such as iChat or Skype. Be warned, however, that despite this appearance, you're still using plain old SMS messages, with each speech bubble in the conversation coming off your monthly message allocation.

Text alerts

When you have unread messages, the SMS icon on the Home Screen will show a small red circle with a digit reflecting the number of new messages. If you also want the iPhone to play an alert sound when you get a text, tap Settings > Sounds and use the New Text Message slider. (The sound won't play if the ringer button is switched to Silent.)

Texting

Clicking the green Messages icon reveals a list of unread messages (signified by a blue dot) and existing "conversations". Tap an entry to view it and you're ready to reply. Alternatively:

- **To delete a message or conversation** Swipe left or right over it to reveal the Delete button. Alternatively, tap Edit, then ⊖.

- **To write a new message** tap ☑ and either enter a phone number, start typing the name of someone in your Contacts list to reveal a list of matching names, or hit ⊕ to browse for a contact.

- **To send a message to multiple people**, start a new message (you can't do this via an existing conversation) and tap ⊕ to add new names. Note that you can't forward SMS messages, but you can copy and paste (see p.61) from an existing text into a new one.

- **To quickly send an SMS to someone in your Favorites or Recents list** tap ⊙ next to their name in the list and choose Text Message.

> **Tip: Street addresses, emails, weblinks or phone numbers in a text conversation can be tapped to launch Maps, Mail or Safari, or to start a call.**

• **To send a photo or video** tap the phone icon next to the text area and either take a picture, shoot a video or point to an image or clip already stored on the iPhone.

• **To call or email someone from your Text Messages list** tap a message in the list, scroll to the top of the conversation and tap Call or, to see their other numbers and email address, click Contact Info.

• **To add someone you've texted to your Contacts list** tap their phone number in the Text Messages list and then tap Add to Contacts.

> **Tip:** **To quickly get to the top of a long SMS conversation and access the Call and Contact Info buttons, tap the time at the top of the iPhone's screen.**

Adding emoticons

If you like using emoticons, also known as emoji icons, in your SMS messages, emails, etc, than download the Emoji app from the App Store, which adds a special custom keyboard set to your iPhone's armoury. Once the app is installed, navigate to Settings > General > Keyboard > International Keyboards > Japanese and enable the Emoji option. From then on, the new keyboard is available via the keyboard switching "globe" button to the left of the spacebar of the iPhone's keyboard.

Chat and messaging

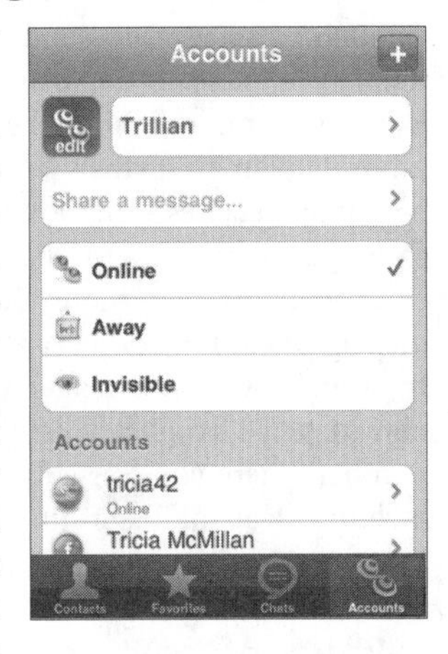

Though the iPhone doesn't offer any built-in tools for instant messaging – also known as chat – you can access all the common services using the device. Once logged in to your network of choice via an app, you can chat with people on computers in addition to those on their phones. It's global, instant and free.

Although it's possible to use chat services via Safari, it's usually far better to download an app designed for the job. Most of the big chat services offer an iPhone app, though if you have contacts spread across many different networks you might prefer a multi-network chat app, such as BeeJive or Trillian (pictured here), both of which allow you to chat simultaneously across a range of networks including AIM, Facebook Chat, Google Talk, iChat, ICQ, MSN and Yahoo!

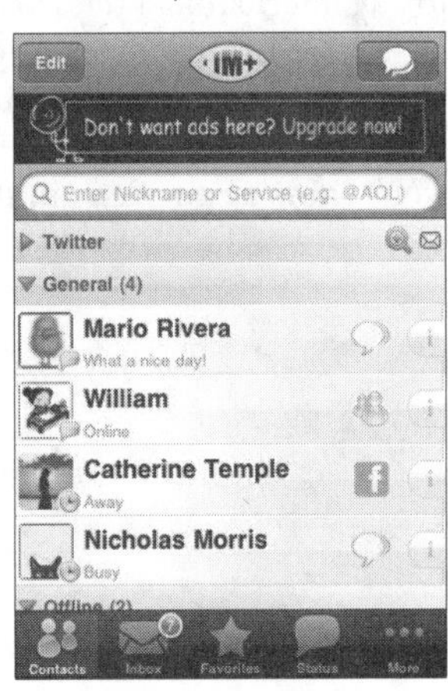

The IM+ app, on the other hand, is another excellent client currently available for the iPhone. It supports live chat with friends and buddies on Skype, Facebook, Google Talk, Yahoo!, MSN, AIM, ICQ, MySpace and Jabber, and can also keep you connected to Twitter.

Texting from your computer

Though we're used to sending SMS messages from our phones, it's also possible to do so from a computer. This can be a great way to save money and make the SMS allocation on your iPhone's monthly plan go further.

Various websites allow you to send free SMS messages – in many cases, go to the carrier website of the person you want to text and enter your message there. But it's also possible to send free text messages – to US numbers at least – from most chat applications. In AOL Instant Messenger or iChat, for instance, just create a new message from the File menu and enter the recipient's mobile number where you'd usually enter a "screen name". You'll need to include +1 at the beginning, as SMS messages sent through the Internet require a country code.

Texting internationally

If you want to text someone in a foreign country, but don't want to pay international rates via your mobile plan, you can use Skype (see p.125) to send texts to most countries for around 8¢/5p. However, you'll want to first change your sender ID to your iPhone number so that people can reply directly. Note that the ID won't show up correctly in the US, China or Taiwan.

iPod

11

iTunes prep

Preparing music and video files to sync with the iPhone

Downloading music and video from the iTunes Store is all well and good, but if you already own the CD or DVD, there's no point in paying for the same content again. "Ripping" CDs and DVDs to get them into iTunes (and in turn onto the iPhone) is easy, but it's worth reading through this chapter even if you've done it hundreds of times, as various preferences and features are easy to miss.

Importing CDs

To get started, insert any audio CD into your Mac or PC. In most cases, within a few seconds you'll find that the artist, track and album names – and maybe more info besides – automatically appear. If your song info fails to materialize (and all you get is "Track 1", "Track 2", etc), or you want to edit what has appeared, either click into individual fields and type, or select multiple tracks and choose File > Get Info to make your changes.

Settings and importing

Before importing all your music, have a look through the various importing options, which you'll find behind the Import Settings button on the General pane of iTunes Preferences. These are worth considering early on, as they relate to sound quality and compatibility. The iPhone can play MP3 and AAC files (up to 320 kbps) as well as Apple Lossless, AIFF, Audible and WAV files (see box opposite).

The bitrate is the amount of data that each second of sound is reduced to. The higher the bitrate, the higher the sound quality, but also the more space the file takes up. The relationship between file size and bitrate is basically proportional, but the same isn't true of sound quality, so a 128 kbps track takes half as much space as a 256 kbps version, but the sound will be only marginally different.

Tip: **To rip multiple tracks as one, simply select them and click Advanced > Join CD Tracks before you import.**

Take a quick look at the other options on offer, but don't worry too much as the defaults will do just fine. When you are ready to import, hit the Import button in the bottom-right corner of the iTunes window.

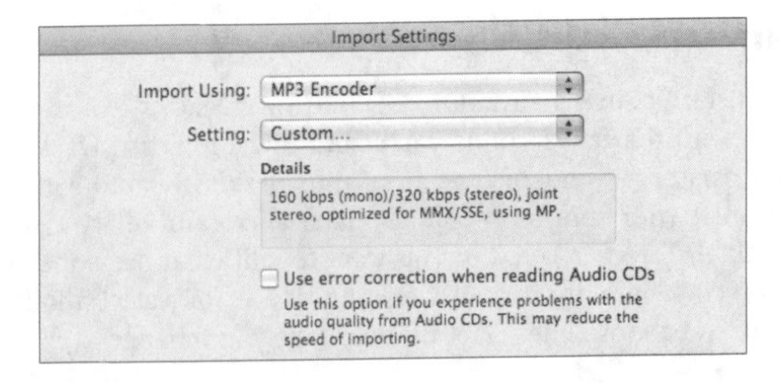

A Rough Guide to music file formats

Music can be saved in various different file formats, just like images (bitmap, jpeg, gif, etc) and text documents (doc, txt, rtf, etc) can be. When you import a CD to iTunes, you can pick from AAC, MP3, Apple Lossless, Wav and Aiff. Here's the lowdown on each:

MP3 [Moving Pictures Experts Group-1/2 Audio Layer 3].
Pros: Compatible with all MP3 players and computer systems. Also allows you to burn high-capacity CDs for playback on MP3-capable CD players.
Cons: Not quite as good as AAC in terms of sound quality per megabyte.
File name ends: .mp3

AAC [Advanced Audio Coding]
Pros: Excellent sound quality for the disk space it takes up.
Cons: Not compatible with much non-Apple hardware or software.
File name ends: .m4a (or .m4p for protected files from the iTunes Store).

Apple Lossless Encoder
Pros: Full CD sound quality in half the disk space of an uncompressed track.
Cons: Files are very large and only play on iTunes, iPhones, iPads and iPods.
File name ends: .ale

AIFF/Wav
Pros: Full CD sound quality. Plays back on any system.
Cons: Huge files.
File name ends: .aiff/.wav

Converting one music file format to another

iTunes allows you to create copies of imported tracks in different file formats. This is great for reducing the size of bulky WAV, AIFF or Apple Lossless files, or for creating MP3 versions of songs that you want to give to friends who have non-Apple music players or phones. Be warned, however, that re-encoding one compressed file format (such as AAC) into another (such as MP3) will damage the sound quality somewhat.

To create a copy of a track in a different format, first specify your desired format and bitrate on the iTunes Import Settings, within the General pane of iTunes Preferences. Then, close Preferences, select the file or files in question in the main iTunes window and choose Create MP3/WAV/AAC Version from the Advanced menu.

When copying high bitrate songs to your iPhone, you can set iTunes to convert them automatically to 128 kbps by checking the appropriate box under the Info tab of your iPhone's options panel in iTunes.

Importing DVDs

Just as with music, before you can transfer video files to your iPhone, you first have to get them into iTunes. In most cases, it's perfectly possible to do this from DVD, though in some countries this may not be strictly legal when it comes to copyrighted movies. As long as you're only importing your own DVDs for your own use, no one is likely to mind. The main problem is that it's a bit of a hassle. A DVD contains so much data that it can take more than an hour to "rip" each movie to your computer in a format that'll work with iTunes and an iPhone. And if the disc contains copy protection, then it's even more of a headache.

Using HandBrake

Of the various free tools available for getting DVDs into iTunes, probably the best is HandBrake, which is available for both Mac and PC. Here's how the process works:

- **Download and install HandBrake** from handbrake.fr

- **Insert the DVD** and, if it starts to play automatically, quit your DVD player program.

DVD copy protection

DVDs are often encrypted, or copy protected, to stop people making copies or ripping the discs to their computers. PC owners can use a program such as AnyDVD (slysoft.com) to get around the protection, while Mac owners trying to get encrypted DVDs into iTunes will need to grab a program such as Fast DVD Copy (fastdvdcopy.com). This allows you to make a non-protected copy of the movie, which you can then turn into an iPad-formatted version using HandBrake. Note that, in some countries, it may not be legal to copy an encrypted DVD.

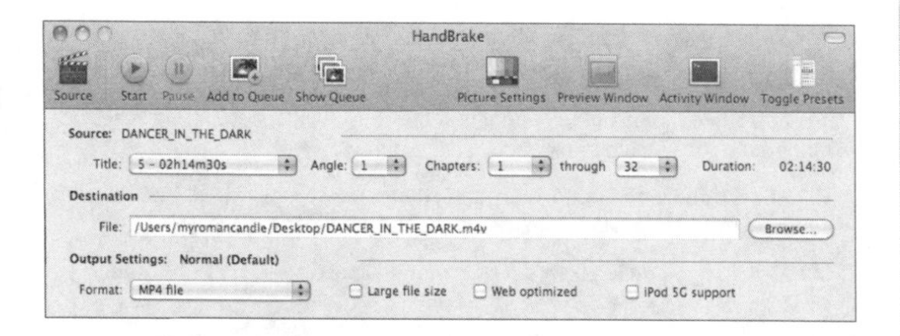

Tip: Ripping Blu-ray discs is at present a prohibitively complex and time-consuming process. Better to hunt out discs that come with an accompanying digital version on DVD.

- **Launch HandBrake** and it should detect the DVD (it may call it something unfriendly like "/dev/rdisk1"). Press Open, and wait until the application has scanned the DVD.

- **Choose iPhone-friendly settings** Choose the iPhone option from the Presets menu.

- **Check the source** It's also worth taking a quick look at the Title dropdown menu within the Source section of HandBrake. Choose the one that represents the largest amount of time (say 01h22m46s) as this should be the main feature. If nothing of an appropriate length appears, then your DVD is copy protected.

- **Subtitles** If it's a foreign-language film, set Dialogue and Subtitles options from the dropdown menus behind the Audio & Subtitles tab.

Tip: Some DVDs feature promotional codes that entitle you to a free iTunes digital version of the same movie.

Supported video formats

To get technical for a moment, note that the various video formats that the iPhone supports include:

H.264-encoded video up to 720p at 30 frames per second, with AAC-LC audio up to 160 kbps, 48kHz, in m4v, mp4 and mov formats.

MPEG-4 up to 640x480 pixels, 2.5 Mbps, with AAC-LC audio up to 160 kbps, 48 kHz.

Motion JPEG (M-JPEG) Avi-wrapped and up to 1280x720 pixels, 35 Mbps, 30 frames per second with ulaw PCM stereo audio. (This is the video format often used by digital cameras that can also shoot video.)

If you find that you have files on your computer that don't meet these criteria, either convert them within iTunes or use HandBrake, as previously described.

- **Rip** Hit the Start button at the bottom of the window and the encoding will begin. Don't hold your breath.

- **Drop the file into iTunes** Unless you choose to save it somewhere else, the file will eventually appear on the Desktop. Drag the file into the main iTunes window. This should create a copy of the new file in your iTunes Library allowing you to then delete the original file from your Desktop.

Converting video files in iTunes

If you find yourself with a video file in iTunes that can't be copied across to the iPhone when you sync, you can easily convert it to an iPhone-friendly format. Select it within iTunes and choose Advanced > Create iPod or iPhone Version.

Tip: **To learn about importing from other music and video sources, see *The Rough Guides to iPods & iTunes*.**

Recording from vinyl or cassette

If you have the time and inclination, it's perfectly possible to import music from analogue sound sources such as vinyl, cassette or radio into iTunes and onto your iPhone. For vinyl, the easiest option is to buy a USB turntable, such as those from Ion or Kam (pictured), but this isn't strictly necessary. With the right cables, you can connect your hi-fi, Walkman, minidisc player or any other source to your computer and do it manually.

• **Hooking up** First of all, you'll need to make the right connection. With any luck, your computer will have a line-in or mic port, probably in the form of a minijack socket (if it doesn't, you can add one with the right USB device; ask in any computer store). On the hi-fi, a headphone socket will suffice, but you'll get a much better "level" from a dedicated line-out.

• **Check that you have enough disk space** During the actual recording process, you'll need plenty of hard-drive space: as much as a gigabyte for an album, or 15MB per minute. Once recording, you can convert the music that you've imported into a space-efficient format such as MP3 or AAC, and delete the giant originals.

• **Choose some software** Recording from an analogue source requires an audio editing application. You may already have something suitable on your computer, but there are also scores of excellent programs available to download off the Internet. Our recommendations are GarageBand for Mac users, which anyone with an Apple machine purchased in the last few years will already have, and Audacity, which is available for both PC and Mac; it's easy to use and totally free:

Audacity audacity.sourceforge.net
GarageBand apple.com/garageband

• **Recording** Connect your computer and hi-fi as described above, and switch your hi-fi's amplifier to "Phono", "Tape" or whichever channel you're recording from. Launch your audio recorder and open a new file. The details from here on vary according to which program you're running and the analogue source you are recording from, but, roughly speaking, the procedure is the same.

You'll be asked to specify a few parameters for the recording. The defaults (usually 44.1kHz, 16-bit stereo) should be fine. Play the loudest section of the record to get an idea of the level. A visual meter should display the sound coming in.

If your level is too low, make sure your line-in volume level is up: on a Mac, look under System Preferences > Sound; on a PC, look in Control Panel.

When you're ready, press "Record" and start your vinyl, cassette or other source playing. When the song or album is finished, press "Stop". Use the "cut" tool to tidy up any extraneous noise or blank space from the beginning and end of the file; fade in and out to hide the "cuts", and, if you like, experiment with any hiss and filters on offer.

• **Drop it into iTunes** When you are happy with what you've got, save the file in WAV or AIFF format, import it into iTunes (choose Import… from the File menu), convert it to AAC or MP3 (see p.122) and delete the bulky original from both your iTunes folder and its original location.

Housekeeping

Playlists

A playlist is a list of songs or videos – a bit like the mixtapes of old. They are not just a fun way to arrange your collection in iTunes, they also provide a useful way to specify which of your music and videos you want to sync with the iPhone (see overleaf). Playlists live at the bottom of the Source List on the left of the iTunes window. There are two ways to create a playlist:

• **Regular playlists** To create a regular playlist, hit the + button at the bottom-left of the iTunes window (alternatively drag one or more items into some blank space in the playlist area or select a bunch of material and choose File > New Playlist from Selection). Once a playlist exists, you can drag in individual songs or add entire albums, artists or genres. You can also safely delete items from the playlist – this won't delete them from your computer, just from that specific playlist.

• **Smart Playlists** Rather than being compiled manually, these are put together by iTunes according to a set of rules that you define. It might be songs with a certain word in their title, or a set of genres, or the tracks you've listened to the most – or a

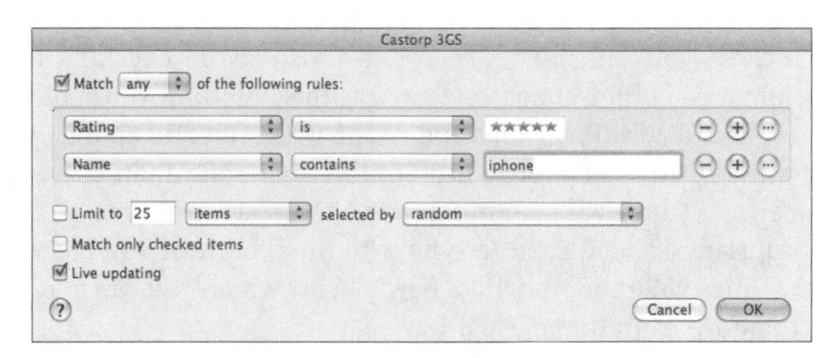

combination of any of these kinds of things. The clever thing is that their contents automatically change over time, as relevant tracks are added to your collection or existing tracks meet the criteria by being, say, rated highly. To create a new Smart Playlist, look in the File menu or click the New Playlist button while holding down Alt (Mac) or Shift (PC).

Editing track info

A common niggle on the iPhone and in iTunes is that you might end up with inconsistent labelling of artists, composers or other track information. For example, you might find that you have one album by Miles Davis and another by Davis, Miles. Thankfully you can quickly edit this information. Simply select one or more tracks – or even whole albums, artists, composers or genres – and choose Get Info from the file menu.

Syncing with the iPhone

Once you have all your audio and video files organized and ready to sync, connect your iPhone to your computer and check the boxes for the content you want to move across within the Music, Movies and TV Shows tabs of iTunes.

Take note of the options at the top of the Music tab. If you expect to use the iPhone's Memos app (see p.227), check the box to include voice memos in your sync; also, avoid checking the "Automatically fill free space…" box, as this will limit your ability to download additional apps and music when out and about.

If you try to load more music onto the iPhone than there is space for, iTunes will ask if you want to create a playlist of the appropriate size and set it to sync with the iPhone. If you answer yes, iTunes will randomly fill a new playlist which you can add to and remove from in the usual way.

Moving music from iPhone to computer

Music purchased on iTunes

If you plug your iPhone into a computer other than the one it's paired with, it will allow you to copy any songs downloaded from the iTunes Store onto the computer. This option may pop up automatically; alternatively, click Transfer Purchases from… in the File menu whenever an iPhone is connected.

Of course, the music will only play back if the computer in question is one of the five machines authorized by your iTunes Store account (see p.152).

Other music

As with iPods, you can't copy music *not* purchased at the iTunes Store from iPhone to computer. This setup is designed to stop people sharing copyrighted music, but can be a real pain if your computer is stolen or destroyed, and the only version of your music collection you have left is the one stored on your iPhone.

Since the iPod and iPhone became popular, many applications have become available allowing iPhone and iPod-to-computer copying. These have never been formally recognized by Apple, but they've generally worked well enough. Two worth investigating are The Little App Factory's iRip and Kennett Net's Music Rescue:

The Little App Factory thelittleappfactory.com
Kennett Net kennettnet.co.uk

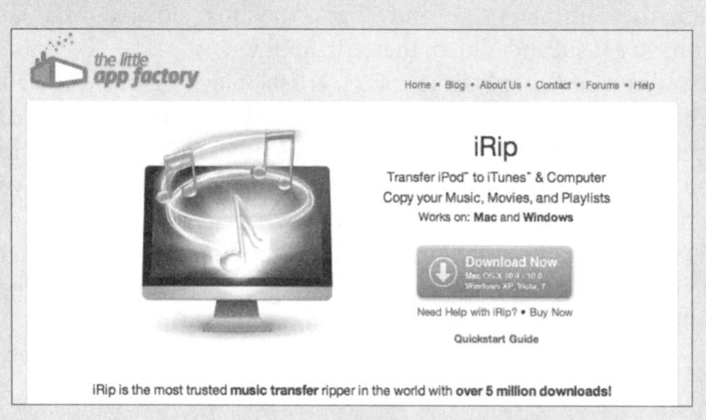

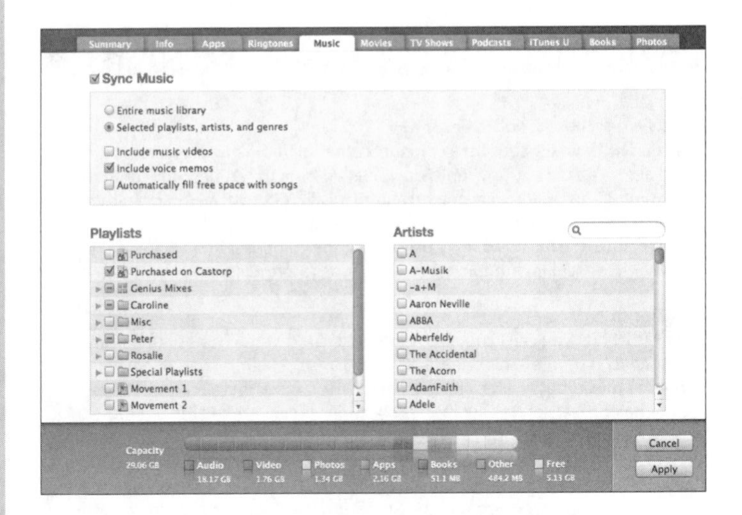

If you prefer, instead of having the iPhone sync with a particular playlist or set of playlists (see p.145), you can add music to your iPhone using simple drag-and-drop. To opt for this approach, click the Summary tab and check the box labelled Manually Manage Music and Video; then hit Apply.

Now you can simply drag tracks, artists, playlists, or even whole genres (if you have the space) straight onto your iPhone's icon from within iTunes. Clicking the triangular icon to the left lets you see the contents of your iPhone in more detail, allowing you to drag music into specific playlists.

With the iPhone set up in this way, removing music is also handled manually, but don't worry, deleting a song from your phone will not affect the original in your computer's iTunes library.

12

The iTunes Store

Buying and renting, direct from the iPhone

The iTunes Store isn't the only option for downloading music and video from the Internet. But if you use the iPhone, it's unquestionably the most convenient, offering instant, legal access to millions of music tracks and music videos, plus a growing selection of TV shows and movies to either buy or rent. Unlike some download stores, the iTunes Store is not a website, so don't expect to reach it with Safari – the only way in is via either iTunes on a Mac or PC, or via the iTunes icon on your iPhone's Home Screen.

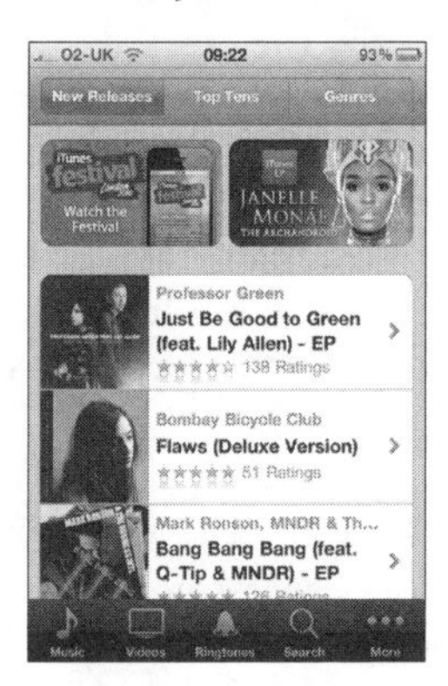

What have they got?

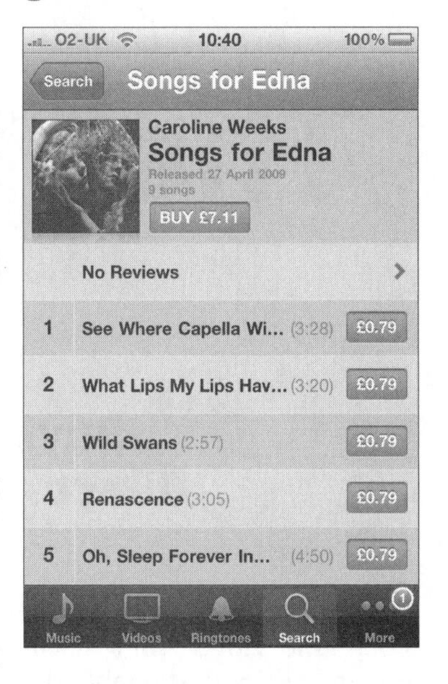

At the time of writing, the iTunes Store boasts more than eleven million tracks worldwide, plus twenty thousand audiobooks and many thousands of movies, TV shows and podcasts (see box). It claims to have the largest legal download catalogue in the world.

However, there are some glaring music and movie omissions, and it isn't like a regular shop where anything can be ordered if you're prepared to wait a while. As with any download site, everything that's up there is the result of a deal struck with the record label or movie company in question, and several independent record distributors have refused to sign up. So don't expect to find everything you want.

That said, thousands of new music tracks, audiobooks, TV shows and feature films appear week after week, so the situation is getting better all the time.

Tip: To access the Store from your computer, click the Store icon in the iTunes sidebar; once downloaded, you can then sync purchased content to the iPhone as and when you need to.

Podcasts

The best way to understand podcasts is to think of them as audio or video blogs. Like regular blogs, they are generally made up of a series of short episodes, or posts, which are nearly always free. Podcasts often consist of either radio-style spoken content or condensed documentary or chatshow-style video episodes, covering everything from current affairs and poetry to cookery and technology. There are many musical podcasts, too, though there's a grey area surrounding the distribution of copyrighted music in this way.

Podcasts are made available as files (either audio or video) that can either be downloaded straight to your iPhone or, alternatively, to a Mac or PC, from which you can sync them across.

Subscribing to podcasts from a Mac or PC

The iTunes Store offers by far the easiest method of subscribing to podcasts. Open iTunes, click iTunes Store in the sidebar and then click the Podcast tab to start browsing or searching for interesting-looking podcasts. When you find one that looks up your street, click Subscribe and iTunes will automatically download the most recent episode to your iTunes Library. (Depending on the podcast, you may also be offered all the previous episodes to download.)

To change how iTunes handles podcasts, click Podcasts in the sidebar and then the Settings button at the bottom. For example, if disk space is at a premium on your system, tell iTunes to only keep unplayed episodes.

To sync your podcasts over to the iPhone, connect it and look for the options under the Podcasts tab.

If you want to stop your kids accessing podcasts through iTunes on your Mac or PC, check the relevant option under the Parental tab of iTunes Preferences.

Subscribing to podcasts on the iPhone

From the Home Screen, tap iTunes > More > Podcasts and browse just as you would any other department of the store. When you find something you want, tap the FREE button to start an episode downloading.

To play podcasts, by default tap iPod > More > Podcasts. For the full story on podcast playback, turn to p.163.

Rolling your own podcasts

Once you've subscribed and listened to a few podcasts you might decide it's time to turn your hand to broadcasting and get in on the action. To find out how to get started, visit apple.com/itunes/podcasts/specs.html

iTunes Accounts

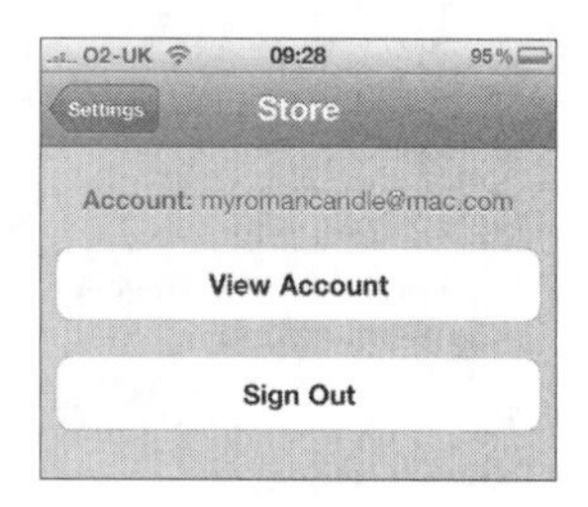

Though anyone can browse the iTunes Store on the iPhone, listen to samples and watch previews and movie trailers, if you actually want to buy anything you need to set up an account and be logged in. If you haven't already done this, it's easily done: either try to buy something, and follow the prompts, or head to Settings > Store and enter the necessary details there.

If someone else is already signed in to the Store on the same iPhone, they'll need to sign out first within Settings > Store. Also note that iTunes Store Accounts are country-specific; in other words, you only get to access the store of the country where the credit card associated with the account has a billing address.

Tip: If you already have an Apple ID, MobileMe Account or iBookstore log-in, exactly the same credentials will work here; you can also use an AOL account if you have one.

Staying secure

After you make a purchase in the iTunes Store and enter your password, the iPhone will remember your details for a short while allowing you to make subsequent purchases. However, there is no way to force the iPhone to remember your iTunes account password permanently, so get used to entering it frequently. For added security, use a passcode (Settings > General > Passcode Lock) to control who has access to your iPhone and how quickly the screen locks after each use.

Renting movies

Many of the major movie studios are now making their films (both new releases and catalogue titles) available for rental via the iTunes Store. Some titles are additionally available in high definition (HD) with a slightly higher rental cost. Once a rented movie file has been downloaded to your iPhone you have 30 days to start watching it, and once you have played even just a few seconds of the file, you have a certain period of time to finish it (in the US it's 24 hours, in the UK, 48). When your time runs out the file miraculously disappears.

A movie rented on the iPhone can't then be transferred to a computer or other device to be watched there. You can, however, rent a movie through iTunes on a Mac or PC and then sync it across to the iPhone.

Annoyingly, you need to fully download a movie before you can start watching it, which, depending on your Internet connection, could take a few hours. To see how your download is progressing, tap More > Downloads at the bottom of the iTunes Store window.

Tip: **If you want to stop your kids either accessing the Store entirely on the iPhone or being exposed to explicit material, look for the options under Settings > General > Restrictions.**

More in the Store

New features are added to the iTunes Store on a regular basis. Here's a list of a few other things that you might like to explore.

• **Redeeming gift cards** If you are lucky enough to have been given one, you can use iTunes Store gift cards and certificates to pay for content in the Store. To redeem a gift, tap Music, scroll down, tap Redeem and follow the prompts. Your store credit then appears with your account info at the bottom of most iTunes Store screens.

• **iTunes U** iTunes U ("university") makes available lectures, debates and presentations from US colleges as audio and video files. The service is free and has made unlikely stars of some of the more entertaining professors.

• **Freebies** Keep an eye open for free tracks: you get something for nothing, and you might discover an artist you never knew you liked. Also, make the most of all the movie trailers on offer in the videos department; simply navigate to a file and tap Preview.

Tip: Purchases sync back to iTunes on your computer when you connect your iPhone; they can be viewed in the Purchased on… playlist that appears in the iTunes sidebar.

DRM and authorized computers

All tracks in the iTunes Store come without traditional built-in DRM (Digital Rights Management) of the kind that Apple used to use to stop content being copied and passed on. Without the built-in DRM there are no technological barriers to someone distributing the files they have purchased. However, files downloaded from the iTunes Store do contain the purchaser's name and email details embedded as "metadata" within the file. The upshot of this is that files purchased from iTunes and then illegally distributed over the Internet are traceable back to the person who originally shelled out for them. The other thing worth noting about iTunes-purchased files is that they're in the AAC format, so they'll only play in iTunes, on iPods, iPhones, iPads and any non-Apple software and hardware that supports this type of file; unless, of course, you convert them to MP3 first (see p.139).

Apple also use what's known as FairPlay DRM, where content purchased using a specific iTunes Account can only be played on (or synced to, in the case of books and apps) a maximum of five Macs or PCs that are "authorized" for that account. (You can, however, add content to as many iPads, iPods and iPhones as you want; you just won't be able to move it onto a Mac or PC that has not been authorized.)

To manage the machines authorized for your iTunes Account, open iTunes on your Mac or PC and look for the options in the Store menu. To deauthorize all machines and start afresh (useful if you no longer have access to one or more of your five), choose Store > View My Account > Deauthorize All.

- **Genius** Tap the Genius tab to get movie, TV show and music recommendations based upon items you've already bought.

- **Ringtones** The store also has a custom ringtones section, which is discussed at length on p.52.

> **Tip:** To change which departments appear on the lower panel of the iTunes Store app on your iPhone, launch the app and tap More > Edit; then drag and drop the various icons to get the configuration you want.

13

iPhone audio

iPod, Remote and other apps

Once your iPhone is loaded up with tunes, you're ready to adjust its on-board audio settings (see box on p.139) and start listening. But there's much more to audio on the iPhone than just the built-in iPod. This chapter will show you a raft of apps that can do everything from turning your iPhone into a remote control to allowing you to tune into your favourite radio station.

The iPhone's iPod

The iPhone's built-in iPod looks great, is easy to use, and is your one-stop shop for playing music, podcasts, audiobooks and music videos. Tap the iPod icon on the Dock to start.

Editing the options

The More button reveals further options for browsing. Depending on your listening habits, these might be more useful than the default options. For instance, classical music buffs will want instant access to Composers, while radio lovers will want to front-load Podcasts. To replace an existing browse icon with a different one, click More, then Edit, and then simply drag the new icon onto the old one. You can also drag the icons at the bottom into any order.

From there on in, it really is pretty self-explanatory: tap to see listings and then tap a track to hear it. You can also tap Now Playing to enter the full-screen "Now Playing" mode.

Now Playing...

Tapping the artwork reveals or hides specific controls, and the ⬅ button takes you back to your listings view. Most of the controls are pretty intuitive and need little elaboration, but for those of you who have just landed from Mars, here's a run-through of what's on offer.

Tip: **While browsing your music collection, any list of songs will include a Shuffle button at the top (pictured). Click to start a random selection of the current list.**

• **Pause/Play a song** Tap **II** and ▶ respectively, or press the mic button on the iPhone headset.

• **To skip** to the start of the current or next song, tap **I◀◀** or **▶▶I**. In a podcast or audiobooks these buttons skip between chapters. You can also skip forward by quickly pressing the mic button on the iPhone headset twice.

• **Scrubbing** To rewind or fast-forward within a song, slide the progress dot on the "scrubber" bar to the left or right. Alternatively, press and hold the **I◀◀** and **▶▶I** controls.

> **Tip:** **Slide your finger up and down the screen to adjust the rate of the scrubbing; this feature works for video playback too.**

• **To see the track list** of all the songs of the current album, tap the ▤ button. Tap the artwork preview, top-right, to get back to the Now Playing screen.

• **To adjust the volume** drag the lower on-screen slider to the left and right, or use the physical buttons on the side of the iPhone.

• **Shuffle** Tap the ⤮ icon to turn the shuffle selection feature on (blue) or off (white).

Cover Flow

A few years ago, Cover Flow was a standalone application, created by independent programmer Jonathan del Strother, that iTunes users could download as an alternative way to browse their music libraries. Apple liked it so much they bought the technology and incorporated it into iTunes, the iPhone, and now even their operating system, OS X Leopard.

Cover Flow is a slick-looking graphical interface for "flicking through" your music's album artwork, so it's perfectly suited to the iPhone, where you really *can* flick. It's like rooting around the record bins of your favourite music store. To switch to the Cover Flow view on the iPhone, simply rotate the device through ninety degrees. The built-in accelerometer recognizes the shift of axis and displays your music library by its artwork. To view the track listing of an album, either tap the relevant image or hit ⓘ. Then tap any of the songs in the list to set it playing.

The only disadvantage of Cover Flow mode is that you don't have access to Shuffle, Repeat or Ratings.

- **Repeat** The iPod app offers two repeat modes, which are available via the 🔁 icon. Tap it once (it goes blue) to play the current selection of songs round and round forever. Tapping it again (🔂) repeats just the current track.

> **Tip:** You can have your iPhone play music, podcasts or videos for a certain amount of time and then switch off – like the sleep function of an alarm clock. Tap Clock and then Timer, and choose a number of minutes or hours. Then tap When Time Ends, choose Sleep iPod, and hit Start.

Music settings on the iPhone

The iPhone offers various options for audio playback. You'll find these by clicking Settings on the Home Screen and then scrolling down to iPod.

• **Shake to shuffle** With this feature turned on, shaking your iPhone while listening to the iPod app will jump you to a random track in the current list (enabling shuffle if it wasn't already on). A source of much confusion, those who don't know if its existence, frequently mistake this feature for a fault with their phone.

• **Sound Check** This feature enables the iPhone to play all tracks at a similar volume level so that none sound either too quiet or too loud. Because these automatic volume adjustments are pulled across from iTunes, the Sound Check feature also has to be enabled within iTunes on your computer. To do this, launch iTunes, open Preferences and under the Playback tab tick the Sound Check box.

• **EQ** Lets you assign an equalizer preset to suit your music and earphones. Note that you can also assign EQ settings to individual tracks in iTunes.

• **Volume Limiter** Lets you put a cap on the volume level of the iPhone's audio playback (including audio from videos), to remove the risk that you might damage your ears or indeed your earphones. Tap Volume Limit and drag the slider to the left or right to adjust the maximum volume level. If you're a parent, you might also want to tap Lock Volume Limit and assign a combination code to prevent your kids from upping the volume level without your permission.

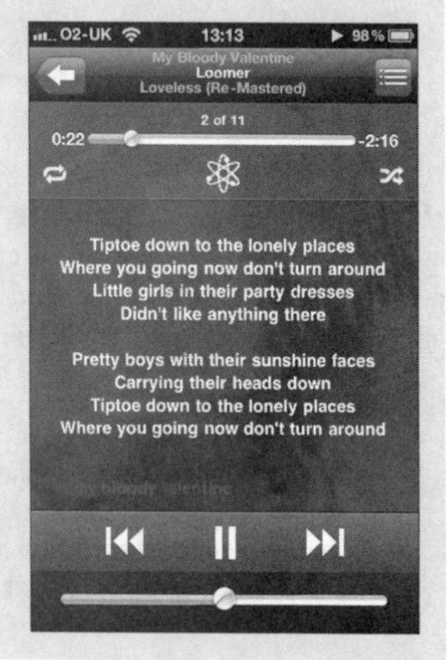

• **Lyrics & Podcast Info** When enabled, the iPod app displays lyric and episode information over the top of the Now Playing artwork screen. This will not work for all files as the data has to be present within the audio file. It generally, however, works for songs downloaded from the iTunes Store.

• **Genius** This feature generates a list of songs from your collection based upon accumulated iTunes Store information. In short, the Genius algorithms recognize that, for example, people who like The Beatles may well also like The Rolling Stones. Tap the ⚛ icon to start – this will base the list on the currently playing song, or, if nothing is playing, prompt you to choose a song to define the list.

• **Create a playlist** Tap Playlists > Add Playlist… and follow the prompts, using the ⊕ icons to add individual songs. To edit playlists, select one to view its contents, tap Edit, and then use the ⊖ icons to remove items, and the draggable ≡ icons to change the order of the list. There are also buttons to Clear the playlist's contents, or Delete it completely.

Tip: Playlists you create on the iPhone will be synced back to iTunes next time you connect.

• **Star ratings** From the "Now Playing" screen, tap the ▤ icon and then slide your finger across the row of dots below the scrubber bar to add a rating for the currently playing selection.

Tip: Sync the music on your iPhone from iTunes using a Smart Playlist of five-star ratings; you can then swiftly remove tracks you don't want by changing their rating on the iPhone.

• **Home button controls** Once music is playing, you can exit the iPod app and continue to listen whilst using other apps. When the screen is locked, double-tapping the Home button reveals the iPod play controls. When using other apps, double-tapping the Home button opens the fast app switching tray (see p.76) where iPod controls can be found by swiping to the right.

Deleting music from an iPhone

As with the iPods, you can't delete unwanted music directly from an iPhone. Instead, simply delete the track in iTunes and it will be deleted from your iPhone next time you connect and sync. Or…

- **If you want the music on iTunes but not on your iPhone**, uncheck the little box next to the names of the offending tracks, and in the iPhone syncing options, choose "Only update checked songs".

- **If you don't want to uncheck the songs**, since this will also stop them playing in iTunes, sync your iPhone with a specific playlist and remove the offending songs from that playlist.

- **If you have Manual Music Management turned on (see p.148)**, simply connect your iPhone to iTunes and browse its contents via the iTunes sidebar (pictured here), deleting songs just as you would from a playlist.

In the case of video podcasts, however, you can delete them directly from the iPhone to free up space – simply swipe across them and then tap Delete to confirm.

The spoken word

If the spoken word is your thing, then it is well worth checking out the Audiobooks and Podcasts sections of the iTunes Store (from either a computer or your phone) to see what's on offer.

Syncing with iTunes is carried out in pretty much the same way as it is for other types of content – you connect your iPhone to iTunes and then check the relevent boxes under the relevant tabs and hit Apply. Podcasts have their own tab, but audiobooks are hidden in the lower section of the Books tab in iTunes and can be easy to miss.

As for listening to your audiobooks and podcasts, the iPod app is the place to go on the iPhone. Playback works the same as with music, although you have a couple of different options available to you from the Now Playing screen:

- **Tell a friend (podcasts only)** Tap the ✉ icon to send an email link to a friend so that they can find the podcast you are listening to in the iTunes Store.

- **Backtrack 30 seconds** Tapping the "30" icon, will rewind the audiobook or podcast by thirty seconds.

- **Playback speed** To adjust the playback speed of an audiobook or audio podcast, tap the "1x" icon to the right of the scrubber bar (to choose either "2x" or "½x").

Remote controls

The iPhone is great as a remote control for any number of audio – and video – setups. Here are a few apps from the App Store that will help you get the job done:

• **Remote** Utilizing the fact that the iPhone can connect to a home Wi-Fi network, this free Apple app turns it into a great control for iTunes on your Mac or PC – to either play directly from the computer, or stream music to networked speakers, connected via an Airport Express base station (apple.com/airportexpress). And for those who have one, the app can also be used to control an Apple TV (apple.com/appletv).

• **Rowmote Pro** Though it'll cost you a few dollars, this super-charged remote control app is well worth having a play with as it gives you access to all sorts of applications on your computer (Macs only at the time of writing), not just iTunes. It's especially useful if you have your computer connected to the television in your living room, allowing you to perform all sorts of tasks from the comfort of your sofa. (Diet and exercise regime not included.)

• **VLC Remote** (pictured) If you use the excellent VLC player on your computer for watching movies, this remote control app is well worth a couple of dollars.

Apps for listening to music

Aside from the iPhone's built-in iPod, there are plenty of other apps that'll let you listen to music … assuming you can connect to the Internet.

• **Spotify** With this app and a Premium Spotify account, you can stream unlimited music to your iPhone from the Internet each month for around the price of a CD album. Though not yet available in the US and many other countries, it's a pretty compelling offering for those who can get it.

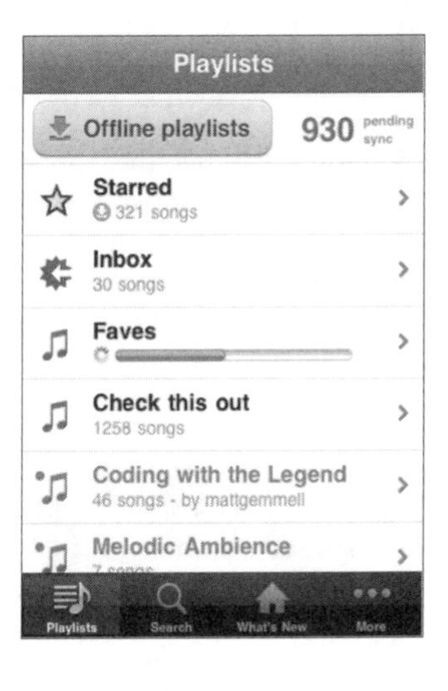

• **AccuRadio** There are plenty of Internet radio players to be found in the App Store; this one is free, has an easy-to-use interface, and lets you create favourites lists.

• **TuneIn Radio** Another great radio tuner app. Though not free, it's still a bargain, given that it offers more than 40,000 stations.

• **Pandora** Only available in the US, this well-established Internet radio service has the added twist of personal recommendations based on your taste.

Tip: **To instantly identify songs using the iPhone's microphone, try the SoundHound or Shazam apps.**

Apps for making music

The Music category within the App Store is awash with virtual instruments, sequencers, drum pads and other noisy creations that, in the hands of most people, will make you want to stuff cotton wool into your ears. With a bit of perseverance and a sprinkling of talent, however, there are some that can be made to sound more than just a novelty. Here's a few worth looking at:

• **GuitarToolKit** When the App Store was first launched the world was wowed by virtual guitar apps. A few years on and there are hundreds available. This is one of the better ones, complete with a tuner and metronome to help you play one of those old-skool wooden versions.

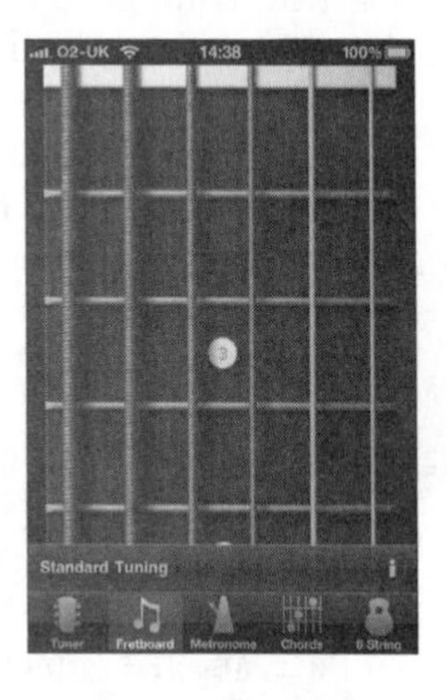

• **Looptastic Producer** An amazing dance-music creation tool, with loads of built-in loops and the ability to add your own. There's also a bunch of effects and time-stretch tools.

• **NLog Free Synth** Arguably the best Korg-styled tone generator for the iPhone, with a delightful analogue feel to the interface.

• **Etude** A massive collection of piano sheet music in a single app, all beautifully rendered on the iPhone's screen. (Getting the thing to balance on a music stand might be a bit of a challenge though.)

Tip: Remember that the headphone jack can be used as a line-out to play your creations via a PA, amplifier or hi-fi system instead of the iPhone's built-in speaker.

14

iPhone video

Movies, TV shows, YouTube

There's nothing particularly revolutionary about a phone or portable media player being able to handle video files. What's different about the iPhone is the size and clarity of its screen, the user-friendly controls and the decent battery life. A new iPhone will play video for around six or seven hours before running out of juice – more than enough for a couple of average feature films.

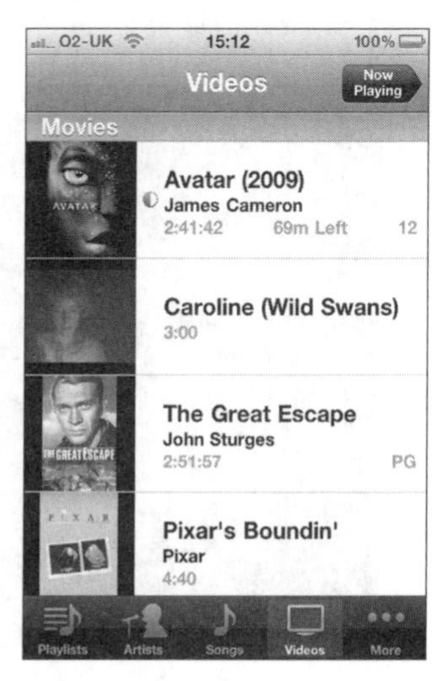

Playing videos

To watch videos on the iPhone, head to the iPod app, tap Videos and start browsing your list. You can also drag to the top of the list to reveal a search field. Depending on the content you have either downloaded from the iTunes Store or synced across from iTunes, you'll find the list divided up into categories: Movies, TV Shows, Podcasts and Music Videos. A blue dot next to a TV Show in the Videos list on the iPhone means that the episode has not yet been viewed. Tap an item to start playback. Once the video is playing, tapping the screen reveals play, volume and scrubbing controls, just like for audio. Additionally, you can:

- **Toggle views** Tap the and buttons icons to toggle between theatrical widescreen and full-screen (cropped) views.

> **Tip:** You can also toggle between the two view modes by simply double-tapping the screen.

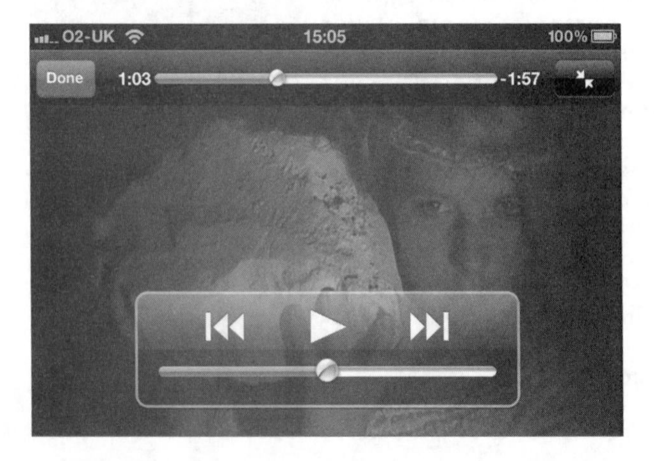

• **Display subtitles** Where the file you are watching supports them, tap the "speech bubble" icon to the left of the play controls to access audio and subtitles options for the playing movie.

• **Display chapters** Where the file you are watching supports them, tap the ▤ icon to the right of the play controls to display the chapters of the movie you are watching.

To delete a movie from the iPhone (which can be really handy if you want to make space for more content when out and about), simply swipe across its entry and then tap Delete to confirm.

Video out

The iPhone can output both NTSC and PAL TV signals (select the one you want within Settings > iPod) so that you can connect your iPhone to a TV or projector. This can be done using one of the following, which all attach to the iPhone via the Dock connector:

• **Apple iPhone Dock Connector to VGA Adapter** This is great for connecting to most standard projectors.

• **Apple Component AV Cable** This gives you five RCA-style plugs: three for the video and two for the audio.

• **Apple Composite AV Cable** This gives you three RCA-style plugs: one for the video and two for the audio.

These accessories need to be purchased separately from Apple. Many third-party manufacturers make cheaper versions, which should work with the iPhone, assuming they have the appropriate Dock connector, though it's always worth checking with the manufacturer before you buy, as some Dock connector cables and adapters may only work with older iPods or iPhones.

YouTube

The iPhone comes with a dedicated YouTube app, which can be used to access all the content from YouTube (assuming you are connected to the Internet). Once a clip is playing, tap to see the on-screen playback controls. They work in exactly the same way as the video controls in the iPod app (see p.168), though here you can additionally:

- **Create Favorites** Tap the 🕮 icon to add a video clip to your Favorites list.

- **Share clips** Tap the ✉ icon to send a link by email.

You can also sign in using a YouTube account log-in (Google account log-ins also work) allowing you to access your uploads, subscriptions, favourites, and also:

- **Leave feedback** Tap the blue ➲ icon next to a clip and scroll down to find the option to Comment or Rate (out of five), or Flag (if you find the clip offensive).

- **Create playlists** Tap More > Playlists to edit, create and delete YouTube playlists.

> **Tip:** If you use YouTube a lot, you might want to edit which icons appear on the category strip at the bottom of the screen. To do this tap More > Edit and then drag and drop the icons you want into position.

Streaming TV

Sure, you can purchase pretty much any TV season you fancy from the iTunes Store, but there are also many streaming TV services that work on the iPhone (even though it lacks Flash). The main limitations to watching either live or catch-up TV on the iPhone are regional: many services that are currently accessible in Europe aren't in the US, and vice versa. You are sure to find some that work, but don't be too surprised if a few of the apps and websites listed below dish you up a whole lot of nothing.

• **TVCatchup** (tvcatchup.com) Pictured here, this is the only thing you need to stream live UK TV to the iPhone for free through Safari. As you might expect, it isn't great over a cellular data connection, but works very well over Wi-Fi.

• **iPlayer** (bbc.co.uk/mobile/iplayer) Until the BBC get around to releasing a dedicated app for viewing their UK catch-up service, you'll have to make do with this Safari-optimized website.

• **At Bat 2010** (app) Live US Major League Baseball with on-screen stats. It's pricey, but if you want to see baseball, it's the best app on the block.

• **TVUplayer** (app) This app pushes out more than 900 live TV channels, but many of them don't display that well on the iPhone. If you stick to those listed with a bandwidth of less than 200 kbps, however, you should be fine.

• **Boxee** (app) Boxee are one of the big names in web-based streaming TV services. Though not released at the time of writing, the Boxee app is rumoured to be in the pipe, so check the App Store, or for their latest news, visit boxee.tv.

Recording from TV

If you want to create iPhone-friendly videos by recording from television, your best bet is to use a TV receiver for your computer. Some high-end PCs have these built in, but if yours doesn't you should be able to pick one up relatively inexpensively, and attach it to your PC or Mac via USB.

The obvious choice for Mac users is Elgato's superb EyeTV range of portable TV receivers, some of which are as small as a box of matches. You can either connect one to a proper TV aerial or, in areas of strong signal, just attach the tiny aerial that comes with the device.

With an EyeTV, it's easy to record TV shows and then export them directly into an iPhone-friendly format. They also produce a very good app that allows you to stream your recordings straight to your iPhone, check the TV schedules and even set recordings going on your EyeTV box at home when out and about. For more info, see:

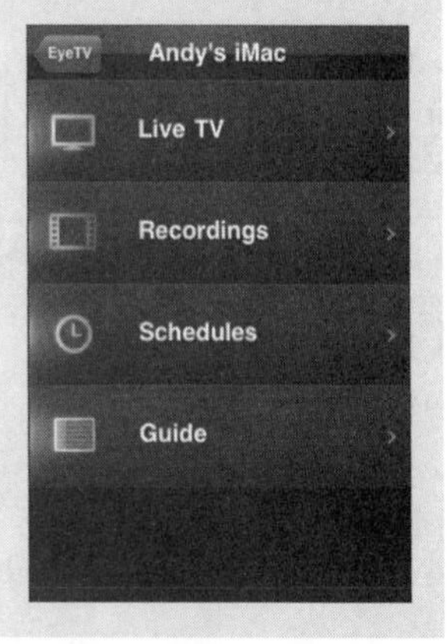

Elgato elgato.com

Hauppauge, Freecom and various other manufacturers produce similar products for PC owners. Browse Amazon or another major technology retailer to see what's on offer. If you can't find one that offers iPhone/iPod export functions, record the shows in any format of your choice and use QuickTime Pro to re-save them for the iPhone.

Internet

15

The web

Safari and beyond

The iPhone certainly isn't the first mobile device to offer web browsing. But, arguably, it's the first one to provide tools that make it a pleasure as opposed to a headache. As discussed earlier, the iPhone comes with a nearly fully fledged version of the Safari web browser. It can't do everything – Flash and Java items won't display, at the time of writing – but it's still one of the most impressive mobile web devices to have ever been produced.

The basics

Make sure you have a phone signal (or even better a Wi-Fi signal), and tap Safari on the Home Screen. Then…

• **Enter an address** Click at the top of the screen, tap ⊗ to clear the current address and start typing. Note the ".com" key for quickly completing addresses (tap and hold it to reveal alternatives).

• **Search Google** Click at the top of the screen, tap in the Google field and start typing. When you're finished, hit Google. If you want to switch from Google to Yahoo! or Bing searching, look within Settings > Safari > Search Engine. Of course, you can also visit any search engine manually and use it in the normal way. For more search tips, see p.182.

> **Tip:** You can see the full URL of any link by tapping and holding the relevant text or image. (This is equivalent to hovering over a link with a mouse and looking at the status bar at the bottom of a normal web browser.)

- **To follow a link** Tap once. If you did it by accident, press ◀.

- **To open a page in a new page** Tap and hold any link (either text or an image) and choose the option from the panel that pops up. For more on using multiple pages, see opposite.

- **Reload/refresh** If a page hasn't loaded properly, or you want to make sure you're viewing the latest version of the page, click ↻.

- **Send an address to a friend** When viewing a page, click at the top of the screen and tap Share. A new email will appear with the link in the body and the webpage's title in the subject line.

History & cache

Like most browsers, Safari on the iPhone stores a list of each website you visit. These allow the iPhone to offer suggestions when you're typing an address but can also be browsed – useful if you need to find a site for the second time but can't remember its address. To browse your history, look at the top of your Bookmarks list, accessible at any time via the 📖 icon. To clear your history, look for the option in Settings > Safari.

Unfortunately, despite storing your history, Safari doesn't "cache" (temporarily save) each page you visit in any useful way. This is a shame, as it means you can't quickly visit a bunch of pages for browsing when you've got no mobile or Wi-Fi reception. It also explains why using the Back button is slower on the iPhone than on a computer – when you click ◀, you download the page in question afresh rather than returning to a cached version.

> **Tip:** **If you struggle with the on-screen keyboard, try rotating the phone when typing in Safari. Unlike Notes and other functions, Safari's keyboard works in both orientations, and the landscape mode offers bigger keys.**

• **Zoom** Double-tap on any part of a page – a column, headline or picture, say – to zoom in on it or zoom back out. Alternatively, "pinch out" with your finger and thumb (or any two digits of your choice). Once zoomed, you can drag the page around with one finger.

Multiple pages

Just like a browser on a Mac or PC, Safari on the iPhone can handle multiple pages at once. These are especially useful when you're struggling with a slow connection, and you don't want to close a page that you may want to come back to later. The only pain is that you can't tap a link and ask it to open in a new window.

• **Open a new page** Tap ⧉, then New Page.

• **Switch between pages** Tap ⧉ and flick left or right. To close a page, tap ⊗.

Bookmarks

Bookmarks, like Home Screen webclips (see p.78) are always handy, but when using a device without a mouse and keyboard, they're even handier than usual. To bookmark a page to return to later on the iPhone, click +. To retrieve a bookmark, tap 📖, browse and then click the relevant entry. To edit your bookmarks:

- **To delete a bookmark or folder** Tap Edit, at the top, followed by the relevant ⊖ icon. Hit delete to confirm.

- **To edit a bookmark or folder** Tap Edit, then hit the relevant entry and type into the name and URL fields.

- **To move a bookmark or folder** Tap Edit and slide it up or down using the ≡ icon. Alternatively, tap Edit, then hit the relevant entry and use the lower field to pick the folder you'd like to move the bookmark or folder into.

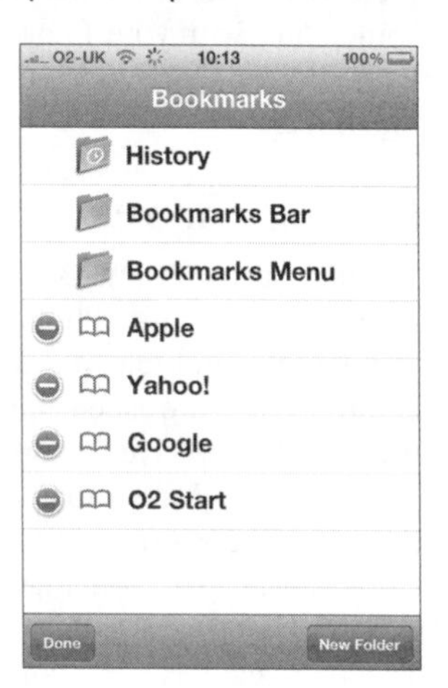

Importing bookmarks from your Mac or PC

iTunes lets you quickly transfer bookmarks from a Mac or PC to your iPhone. Just connect your iPhone and click its icon in iTunes. Under the Info tab, check the relevant box under web Browser. The bookmarks will move across to the iPhone, though they can be easy to miss: if you use Safari on your Mac or PC, you'll find them within two folders labelled Bookmarks Menu and Bookmarks Toolbar.

Likewise, bookmarks from the iPhone will appear back in your browser. In Safari, you won't find them in the Bookmarks menu, however: you'll have to click Show All Bookmarks or the 📖 icon.

You can also choose to sync bookmarks to your iPhone over the airwaves if you have a MobileMe account; look for the option within your specific account settings within Settings > Mail, Contacts, Calendars

> **Tip:** **Use the free Xmarks service (xmarks.com) to sync other browsers' bookmarks to Safari, and in turn the iPhone.**

Forms & AutoFill

You can set Safari on the iPhone to remember the names and passwords that you use frequently on websites, though if you do, there is a risk that someone else could use those credentials on a website if they got hold of your phone. You can turn the feature on or off within Settings > Safari > AutoFill – look for the option to turn on Names & Passwords. If you do enable it, also consider setting a four-digit security code to unlock the phone's Lock Screen, as an extra level of security. This is found within Settings > General > Passcode Lock.

AutoFill can also help you when filling out address fields on webpages, though you will need to tell the iPhone which contact details to use for this. First enable AutoFill, as already described, and then turn on Use Contact Info, choose My Info and select your contact entry from your iPhone's Contacts list.

If at any time you wish to remove all the saved passwords and usernames on your iPhone, tap Clear All.

> **Tip:** **When typing into form fields on a webpage, use the Previous, Next and AutoFill buttons just above the keyboard to make best use of the AutoFill feature.**

Safari settings

You'll find various browsing preferences under Settings > Safari. Here you can empty your cache and history (useful if Safari keeps crashing, or you want to hide your tracks), play with security settings (make sure that the Fraud Warning option is switched on), choose a search engine, turn the Bookmarks Bar on and off, and also access options for the following:

- **JavaScript** This is a ubiquitous way to add extra functions to websites and is best left on.

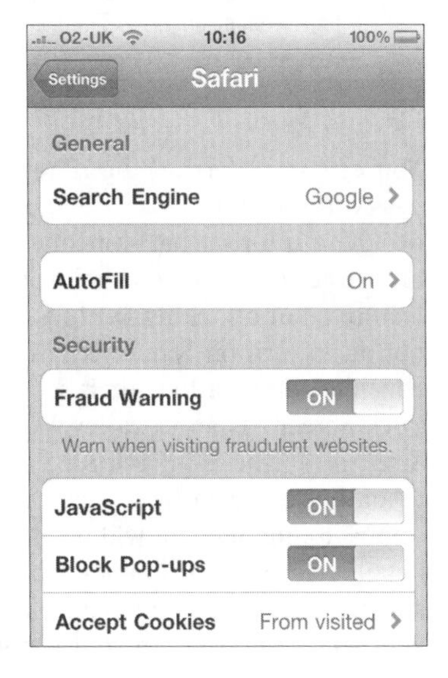

- **Pop-up blocker** stops pop-up pages (mainly ads) from opening.

- **Cookies** These are files that websites save on your iPhone to enable content and preferences tailored for you … for example, specific recommendations on a shopping site.

- **Databases** Some advanced browser-based applications (such as those employed by Google) will store application information on the iPhone in the form of a "database"; these are sometimes also referred to as "super cookies", and allow you to use such applications within a browser when offline. If you are having problems with specific sites within Safari, you may well find that deleting any databases they use can help.

Other browsing tips

Searching for text on a page

One very useful tool currently missing in the iPhone version of Safari is the ability to search for text within a page. This is especially annoying when you follow a link in Google and end up on a very long page with no idea where your search terms appear.

A workaround for this is to get into the habit, when searching the web, of visiting Google's "cached" (saved and indexed) version of a webpage rather than the "live" version. You'll find a link for this option next to each of the results when you search Google. Click "Cached" and see a version of the target page that's got each of your search terms highlighted in a different colour throughout – almost as good as being able to click Find.

Rough Guides Travel
travel and music guide publishers; includes an online guide to destinations throughout the world,...
www.roughguides.com/ - 24k - Cached - Similar pages

Webpage display problems

If a webpage looks weird on screen – bad spacing, images overlapping, etc – there are two likely causes. First, it could be that the page isn't properly "Web compliant". That is, it looked OK on the browser the designer tested it on (Internet Explorer, for example), but not on other browsers (such as Safari or Firefox). The solution is to try viewing the page through another browser.

Second, it could be that the page includes elements based on technologies that the iPhone doesn't yet handle, such as Flash, Real and Windows Media. This is especially likely to be the problem if there's a gap in an otherwise normal page.

If a webpage looks OK, but different from the version you're used to seeing on your Mac or PC, it could be that the website in question has been set up to detect your browser and automatically offer you a small-screen version.

iPhone Googling Tips

The Google search field at the top of each Safari page on the iPhone is an incredibly useful tool and as you start to type, its dynamic popover list of suggested searches makes it even more so. But you can make it work even harder.

Google also offer a very easy to use iPhone-optimized results page, with tabs at the top for jumping between different types of search – "web", "images", "local", etc.

Given that your connection speed may be low when out and about, it makes sense to hone your search skills for use on the iPhone. All the following tricks work on a PC or Mac, too. Typing the text as shown here in bold will yield the following search results:

Basic searches

william lawes > the terms "william" and "lawes"

"william lawes" > the phrase "william lawes"

william OR lawes > either "william" or "lawes" or both

william -lawes > "william" but not "lawes".

All these commands can be mixed and doubled up. Hence:

"william lawes" OR "will lawes" -composer > either version of the name but not the word "composer".

Synonyms

~mac > "mac" and related words, such as "Apple" and "Macintosh".

Definitions

define:calabash > definitions from various sources for the word "calabash". You can also get definitions of a search term by clicking the definitions link at the right-hand end of the top blue strip on the results page.

Flexible phrases

"william * lawes" > "william john lawes", etc, as well as just "william lawes".

Search within a specific site

site:bbc.co.uk "jimmy white" > pages containing Jimmy White's name within the BBC website. This is often far more effective than using a site's internal search.

Search web addresses

"arms exports" inurl:gov > the phrase "arms exports" in the webpages with the term gov in the address (i.e. government websites).

Search page titles

train bristol in title:timetable > pages with "timetable" in their titles, and "train" and "bristol" anywhere in the page.

Number and price ranges

1972..1975 "snooker champions" > the term "snooker champions" and any number (or date) in the range 1972–1975.

$15..$30 "snooker cue" > the term "snooker cue" and any price in the range $15–30.

Search specific file types

filetype:pdf climate change statistics > would find PDF documents (likely to be more "serious" reports than webpages) containing the terms "climate", "change" and "statistics".

Viewing online PDFs and Word documents

The iPhone can view Word, Excel and PDF documents on the web, creating an iPhone-friendly preview version in a popover window. Once opened, scroll down to read subsequent pages, and double-tap to zoom in just as you would with a regular webpage.

In the case of PDFs, and assuming you have the free Apple iBooks app installed (see p.215) also note the link to Open in iBooks. This link saves a copy of the PDF to your iBooks bookshelf. The file is then synced back to iTunes next time you connect.

> **Tip:** If you're following a link from Google to a PDF, Word or Excel doc, and it is taking an age to download, click the "View As HTML" link instead of the main link to the document. This way you'll get a faster-loading text-only version.

Webpages specially optimized for the iPhone

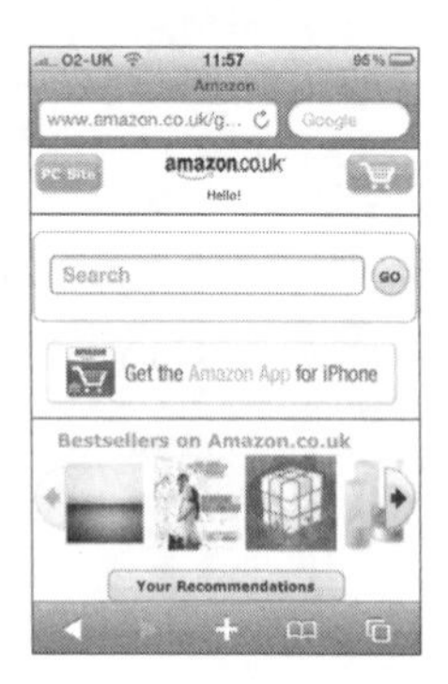

Though the iPhone can handle almost all webpages, and most look great, some are specially designed to work perfectly on the iPhone screen – even mimicking the visual language of columns and buttons found in iPhone apps. Many big name sites, such as eBay, Amazon, Facebook and Gmail, offer such mobile versions of their regular sites. If you like them, use the iPhone's webclip feature (see p.78) to add a shortcut to your Home Screen.

Web apps

Though App Store apps are discussed at length throughout these pages, it's also worth knowing that there are many free-to-use web-based apps and tools (aka web apps) that can be accessed through your iPhone wherever you have an Internet connection. To find out more, turn to p.185.

Browser apps

There are several alternative browsers available in the iTunes App Store, but none that offers all the features and integration of Safari. Browse or search the Utilities and Productivity categories to see what's available, or try one of these:

- **Opera Mini** From the same people who make the brilliant desktop Opera browser, this iPhone version is impressively fast and features tabbed browsing, password tools and the quick access "Speed Dial" tool.

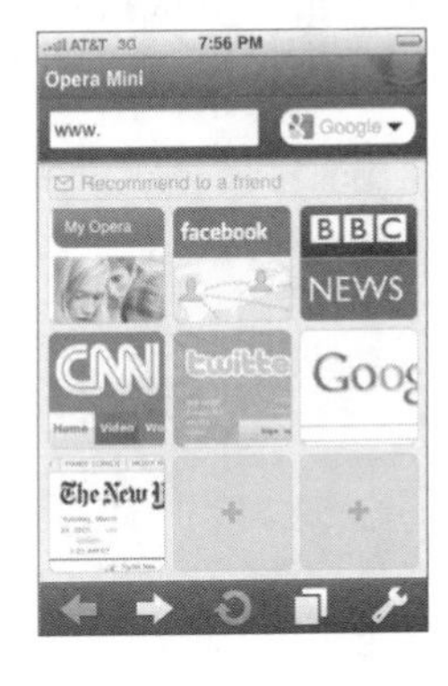

- **FREE Full Screen Private Browsing for iPhone & iPad** A no-frills affair that offers completely private browsing sessions: no history, cache, cookies etc. Catchy app name too.

- **Tabs** If you are not satisfied by Safari's Multiple Pages feature, this is the app to download for a tabbed browsing fix.

- **Quicky Browser** Very handy app for downloading and saving entire webpages so that you can read them offline without an Internet connection.

Site-specific apps

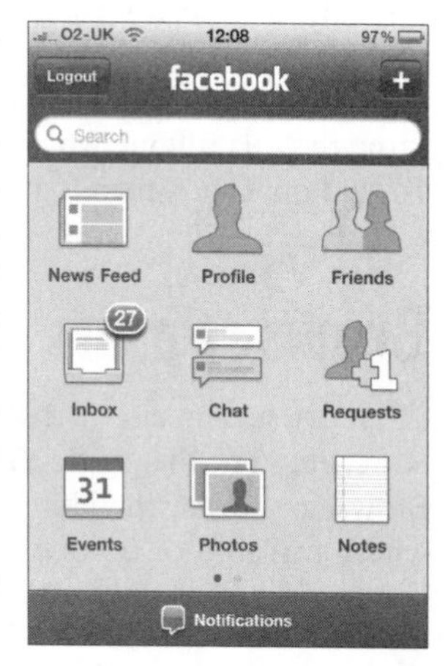

Many popular websites also offer their services in the form of fully fledged apps (see, for example, the free Facebook app pictured here), negating the need to browse to them via Safari all together. The advantage of such apps is that you're likely to have a far more polished and speedy interface at your disposal and, in many instances, some features are made available offline that would otherwise be unusable in a browser. Search for your favourite sites in the App Store to see what's offered.

16

Email

How to set up and use Mail

Having email available wherever you are completely changes your relationship with it. It becomes more like text messaging – but much better. Like recent Macs, the iPhone comes with an email program known as Mail. However, the version on the iPhone has been cut down to its bare bones.

Setting up email accounts

The iPhone's Mail application and Settings panel come pre-configured to work with email accounts from Microsoft Exchange, AOL, Gmail, MobileMe and Yahoo! If you use one of these email providers, you may be used to logging in via a website (and indeed, it's perfectly possible to do this on the iPhone, via Safari). However, it's generally faster and more convenient to use a dedicated mail program – such as Mail on the iPhone.

To set up one of these accounts, just tap Mail (or tap through from Settings > Mail, Contacts, Calendars > Add Account…), choose your account provider from the list and enter your normal log-in details. Under Description, give your email account a label, for example "Personal" or "Work" – this is your means of distinguishing between multiple email accounts within the iPhone's Mail application.

You may be prompted to log in to your account on the web and enable either POP3 or IMAP access (see our jargon buster, below). You can do this via Safari on your iPhone, or using a computer.

Email jargon buster

Email can be collected and sent in various ways, the most common being POP, IMAP and Exchange – all of which are supported by the iPhone. If you're using an account from your ISP, you may find you can choose between IMAP and POP. Here's the lowdown on each type:

• **POP** (or POP3) email accounts can be sent and received via an email program such as Mail or Outlook. Each time you check your mail, new messages are downloaded from your provider's mail server onto your computer or phone. It's a bit like a real-world postal service – and, indeed, POP stands for Post Office Protocol. When using a computer, messages are usually deleted from the server as you download them, but it is possible to leave copies on the server so you can download them from other computers. By default, the iPhone doesn't delete the messages as it downloads them.

• **IMAP** An IMAP account can also be sent and received via an email program, but all the messages are based on your mail provider's server, not on your phone or computer. When you open your mail program, it downloads the email headers (from, to, subject, etc). Clicking on a message will download the body, but not delete it from the server. This can be a bit slow, but it means your mail archive will be available at all times. It also means the amount of mail you can store will be limited by the server space offered by your provider. IMAP stands for Internet Message Access Protocol.

• **Microsoft Exchange** is Microsoft's corporate system. If you use Outlook at work, it's likely that you're using an Exchange email account.

• **Web access** Most POP, IMAP and Exchange email providers also let you send and receive email via a website. You can access your mail this way on the iPhone via Safari, though it's much more convenient to use Mail.

Contacts and calendars too?

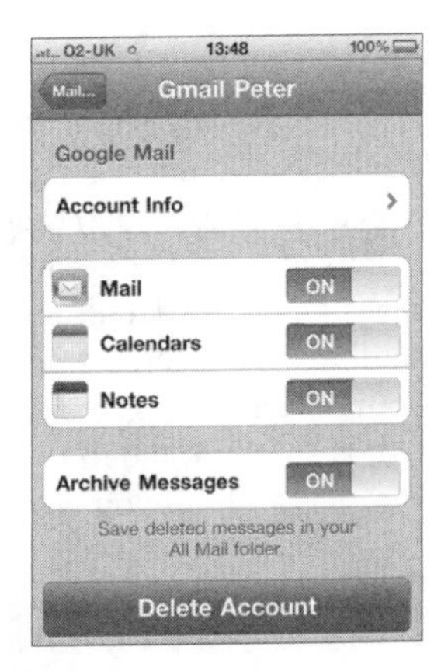

In the case of Microsoft Exchange, Google and MobileMe, you might well be prompted to set up either contacts, calendars or Notes too. This setting will override any previous contact or calendar syncs that you have set up for your iPhone within iTunes. Now your contacts, calendars and notes will be "pushed" from the Exchange, Google or MobileMe server online.

Tip: Both Exchange and Gmail accounts support calendar invitations, which can be accepted or declined via an invitations tray in the iPhone's Calendar app. The number of unanswered invitations shows up as a badge on the app's Home Screen icon.

Setting up corporate Exchange email

Corporate-style Exchange email is fully compatible with the iPhone, though only if the network administrator activates IMAP access on the server. If your office won't allow this, one workaround is to ask your Exchange administrator to allow you to sign up for the SyncYourMail service (pimmanager.divido.dk), which cleverly uses Exchange web access to map your email and other data onto the iPad's Mail application. It costs €40 a year.

Tip: To comply with corporate security protocols, you may well find that as part of the set-up process of an Exchange account you are required to set a Passcode Lock (see p.55) for your iPhone.

Setting up other email accounts via iTunes

If you have a regular email account that you use with Mail, Outlook or Outlook Express (an account from your broadband provider, for example) then it's simple to copy across your account details onto your iPhone. This won't copy across the actual messages – just the details about your accounts so that you can begin sending and receiving on the phone. To get things going, connect your iPhone to your Mac or PC, select its icon in iTunes and choose the Info tab in the main panel. Scroll down, check the box for each account you want to copy across, and press Apply.

Push versus fetch

Traditionally, a computer or phone only receives new emails when its mail application contacts the relevant server and checks for new messages. On a computer, this happens automatically every few minutes – and whenever you click the Check Mail or Send/Receive button. This is referred to by Apple as a "fetch" setup.

By contrast, email accounts that support the "push" system feed messages to the iPhone the moment they arrive on the server – which is usually just seconds after your correspondent clicks the Send button. The Yahoo!, Microsoft Exchange and MobileMe services are all examples that use the push system for emails, contacts and calendars.

To turn on push services for those accounts that support it, tap Settings > Mail, Contacts, Calendars > Fetch New Data. Here, you can also set up how frequently you would like accounts that use fetch to check in with the server to see if there are any new emails available. It's worth noting that the more frequently that emails and other data are fetched, the quicker your battery will run down. These settings also apply to other apps (such as some to-do list tools and instant messaging clients) that rely on Apple's push services to grab your up-to-date data from a server.

If you want the convenience of push email without switching accounts, one option is to set your existing account to forward all your mail to a free Yahoo! account, and set up both on the iPhone. The Yahoo! account will alert you to new messages instantaneously, which will then appear in your regular account when you open Mail, allowing you to reply as normal. Ask your email provider whether they can activate forwarding for you.

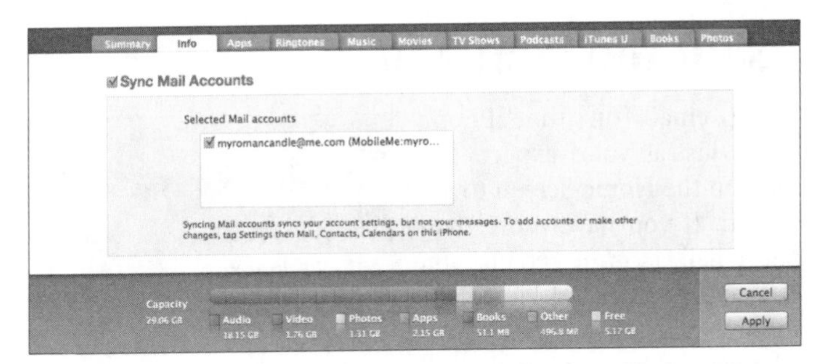

Setting up other email accounts on the iPhone

If you have an email account that you only access via the web, rather than with an email program, and it isn't provided by AOL, Gmail, MobileMe or Yahoo!, then contact your provider to ask whether the account offers POP or IMAP access. If it does, get the details and set up the account on the iPad manually. You should be able to find a page on their website that spells out everything you need to know to set things up. You just need to plug in the account details yourself, being careful to type in everything exactly as your email provider directs you to.

To get started, tap Settings > Mail, Contact, Calendars > Add Account… > Other. Next choose from IMAP, POP or Exchange. If you're not sure, try POP.

Next, fill in the details. If you're not sure and your email address is, say, joebloggs@myisp.com, the username may be joebloggs (or your full email address), the incoming server may be mail.myisp.com or pop.myisp.com; and your outgoing server may be smtp.myisp.com. Press Save when you're done.

Tip: **The Add Account > Other screen can also be used to set up LDAP and CardDAV contacts accounts, and also CalDAV and subscriber calendar accounts.**

Using the Mail app

Using email on the iPhone works just as you'd expect. Tap Mail on the Home Screen to get started. If you have more than one email account set up, you can choose to either view each Inbox separately, or view them all together by tapping the All Inboxes option. You can also…

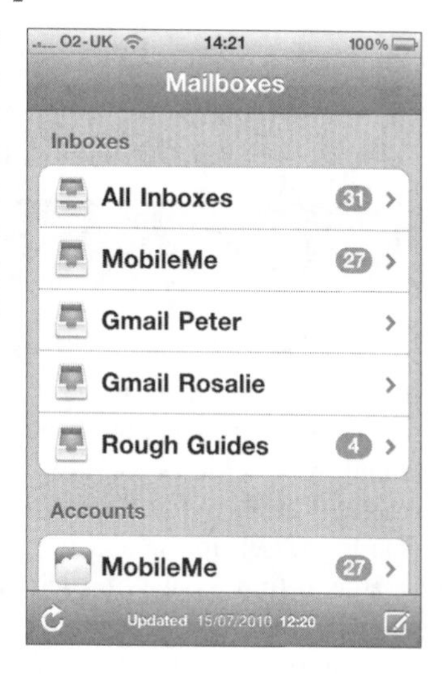

• **Compose a message** Tap ✎. (If you have more than one account set up, first select the account you want to use from the list.) Alternatively, you can kick-start a message by tapping a name in Contacts, Recent or SMS and then tapping the contact's email address.

• **View a message** Tap any email listed in your Inbox to view the entire message. Double-tap and "pinch" respectively – just like with Safari. If you often find that you have to zoom in to read the text, try raising the minimum text size under Settings > Mail.

• **Move between messages** Use the ▲ and ▼ buttons at the top to move up and down through your emails.

• **Move between accounts and folders** Use the left-pointing arrow button at the top (which displays the name of the item one layer up in the hierarchy) to navigate through all your folders and accounts, with the latter farthest to the left.

> **Tip:** Move to the top of long emails or scrolling lists by tapping the iPhone's Status Bar at the top of the screen.

• **Open an attachment** You can open Word, Excel, PowerPoint and iWork files attached to emails. You can also view images and PDFs from emails and save them to Photos and iBooks, respectively, to be synced across to your computer. To save an image, tap it and choose Save Photo. To save a PDF tap the Open in iBooks button.

• **Reply or forward** Open a message and tap ↰.

• **Moving messages** To move one or more messages to a different folder, hit Edit, check the messages and tap Move. Or, when viewing an individual message, tap the button.

• **To attach a photo** You can't add an attachment to a message that you have already started directly from Mail, but you can Paste images that have already been copied using the Copy command in another application. To reveal the Paste command, tap and hold within the message you are composing. You can also create an email from images in the iPad's Photos app: select an image, tap , and follow the prompts. To send multiple images from Photos whilst viewing a grid of images (either an Album, Places or Faces set), tap , then select the images you want to attach to an email and then hit the Share button. If

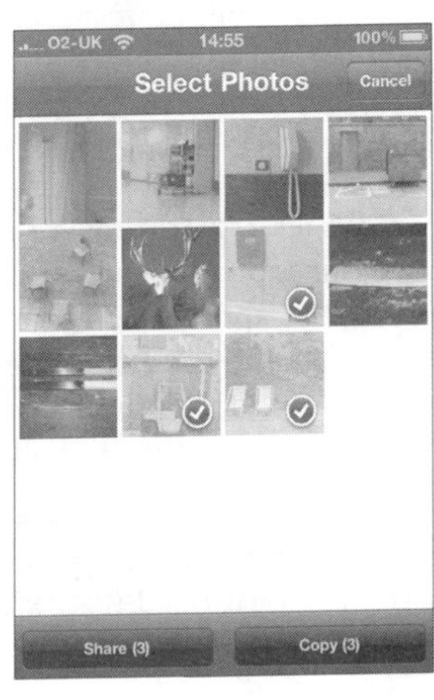

you have multiple email accounts, messages that originate from Photos will be sent from the default account, which you can select from Settings > Mail, Contacts, Calendars.

• **Deleting messages** You can delete a message from a list by swiping left or right over it and then tapping Delete. To delete multiple messages simultaneously, tap Edit and check each of the messages you want to trash. Then click the Delete button.

> **Tip: A third way to delete messages is to open them and press . This then jumps to the next message. If you don't want to waste time confirming each time you hit delete, turn off the Settings > Mail > Ask Before Deleting option.**

• **Empty the Trash** Each email account offers a Trash folder alongside the Inbox, Drafts and Sent folders. When viewing the contents of the Trash, you can tap Edit and either permanently delete individual items or choose to Delete All. Alternatively, tap Settings > Mail, Contacts, Calendars, choose an account, and then tap Account Info > Advanced > Remove and set to have messages in the Trash automatically deleted either never, or after a day, a week, or a month.

• **Create new contacts from an email** The iPhone automatically recognizes phone numbers, as well as email and postal addresses when they appear in an email. Simply press and hold the relevant text to see the options to Create a new contact, Add to an existing contact or Copy to the clipboard. In the case of a postal address, you will also get the option to see the location in Google Maps.

> **Tip: As with Safari, you can tap and hold a link in an email to reveal the full destination address. Useful for links in emails that seem a bit dodgy.**

Tweaking the settings

Once your email account is up and running on your phone, scan through the Settings options to see what suits you. Some things to consider:

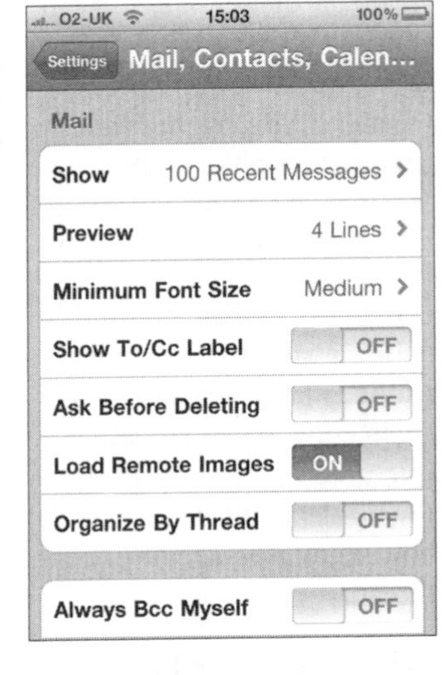

• **Message preview** If you'd like to be able to see more of each message without clicking it, press Settings > Mail, Contacts, Calendars > Preview and increase the number of lines.

• **To/Cc** If you'd like to be able to see at a glance whether you were included in the To: or Cc: field of an email, tap Settings > Mail, Contacts, Calendars > Show To/Cc Label. A small icon will appear by each message preview stating "to" or "cc".

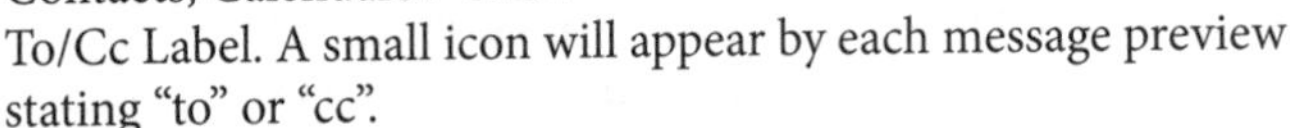

• **Sent mail** With a standard email account, messages sent from your iPhone won't get transferred to the Sent folder on your Mac or PC. If this bothers you, as you'd like to have a complete archive of your mail on your computer, turn on Always Bcc Myself under Settings > Mail, Contacts, Calendars. The downside is that every message you send will pop up in your iPhone inbox a few minutes later. The upside is that you'll get a copy of your sent messages next time you check your mail on your Mac and PC. You can copy these into your Sent folder manually, or set up a rule or filter to do it automatically.

- **Default account** If you have more than one email account set up on the iPhone, you can choose one to be the default account. This will be used whenever you create messages from other applications – such as when you email a picture from within Photos (see p.193).

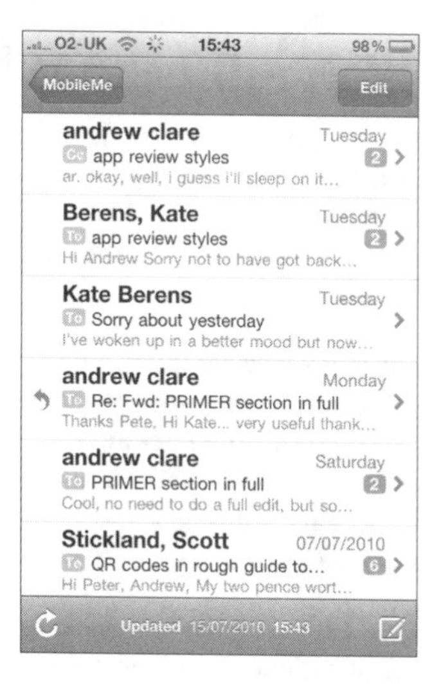

- **Organize by threads** With this option enabled, Mail groups all the email exchanges in a given conversation together (pictured), with the number on the right-hand side displaying how many messages are in the thread.

- **Adding a signature** Even if you have a sign-off signature (name, contact details, etc) set up at home, it won't show up automatically when you use the same account from the iPhone. To set up a mail signature for your iPhone, tap Settings > Mail, Contacts, Calendars > Signature and then enter your signature.

- **Mail days** Microsoft Exchange users can also increase or decrease the number of email messages displayed on the iPhone from an Exchange mail account. This can be done from the "Mail days to sync" option within the specific settings screen for the given Exchange account.

- **Show** Use this option to determine how many messages are displayed within your Inbox… this is particularly useful if you are in denial about the amount of mail you have to get through!

Email problems

You can receive but not send

If you're using an account from your Internet Service Provider, you might find that you can receive emails on the iPhone but not send them. If you entered the details manually on the iPhone, go back and check that you inputted the outgoing mail server details correctly, and that your log-in details are right.

If that doesn't work, contact your ISP and ask them if they have an outgoing server address that can be accessed from anywhere, or if they can recommend a "port" for mobile access. If they can, add this number, after a colon, onto the name of your outgoing mail server – which you'll find by tapping Settings > Mail, Contacts, Calendars, choosing your account, and then tapping Account Info > SMTP. For example, if your server is smtp.myisp.com and the port number is 138, enter smtp.att.yahoo.com:138

If you have more than one email account set up on your iPhone, also try using an alternate outgoing mail server. To make sure that they are all available to be used by any given account, select that account, as above, tap SMTP and then toggle the other servers On within the Other SMTP Servers area of the panel.

There are messages missing

The most likely answer is that you downloaded them to your Mac or PC before your iPhone had a chance to do so. Most email programs are set up to delete messages from the server once they've successfully downloaded them. However, it's easy to change this.

First, open your mail program and view the account settings. In most programs, look under the Tools menu. Click the relevant account and look for a "delete from server" option, which is usually buried under Advanced.

Choose to have your program delete the messages one week after downloading them. This way your iPhone will have time to download each message before they get deleted. It also means you'll have access to more of your messages when checking your mail via the web.

Messages don't arrive unless I check for them

Unless you use an account that supports the "push" mail system, then you are relying on your iPhone's "fetch" settings to retrieve emails for you from the server. To check that it is configured in the way that you want it, tap your way through to Settings > Mail, Contacts, Calendars > Fetch New Data. Of course, your phone will need to have either carrier or Wi-Fi reception to check in with your account.

I get a copy of all the messages I send

It could be that Always CC Myself is switched on under Settings > Mail. But if you're using Gmail the most likely cause is that you have Use Recent Mode switched on. To turn this off, tap Settings > Mail and select your Gmail account; tap Advanced and you'll see the relevant slider at the bottom.

More...

17

Camera & photos

Pictures in your pocket

The iPhone serves both as a stills and video camera in itself and as a photo album to show off your digital photo collection – including pics taken with other cameras. This chapter offers some tips on using the iPhone's built-in camera before explaining how to get your existing images onto your phone.

The iPhone camera

All iPhones have a built-in digital camera, and with each new model the spec and capabilities of this tool have increased. The iPhone 4 has a 5-megapixel stills camera on the back that is also capable of shooting HD video. The phone also boasts a second front-facing lens that can be used for stills, video and Apple's much-hyped FaceTime calls (see p.120).

Shooting stills

To take a still picture, simply tap the Camera button on the Home Screen, aim and tap 📷. Depending which model you have, try some, or all, of the following:

• **Hold it steady** The iPhone takes much better pictures when it's held steady, and when the subject of the picture is not moving. Try leaning on a wall, or putting both elbows on a table with the iPhone in both hands, to limit wobble.

• **Stay focused** The iPhone 4 can auto-focus. To choose exactly which object or person you want the camera to focus on, tap on the screen to move the square focus target around.

• **Shoot outside** The iPhone takes much better images outside, in daylight, than it does inside or at night. However, too much direct light and the contrast levels hit the extremes.

• **Light source** Make sure that the light source is behind the camera and not behind the subject of the photograph.

Tip: Don't forget that you can rotate the iPhone to shoot both still shots and video in landscape mode too.

The iPhone microscope

If you want to get really close in on your subject, consider combining your iPhone's camera with a magnifying lens or microscope. You can then go and make some new friends in this Flickr group: flickr.com/groups/424440@N23

- **Use the flash** If you have an iPhone 4, tap the ⚡ icon, top left, to access the device's built-in flash controls. If you do take shots inside, expect to get better results in light that tends towards white, as opposed to that produced by more yellowy bulbs.

- **Get close and zoom** To take a decent portrait, you'll need to be within a couple of feet of the subject's face. This way more pixels are devoted to face rather than background; in addition, the exposure settings are more likely to be correct. Tap the screen to reveal the zoom slider for additional control.

- **Self portraits** tap the "camera switch" button, top-right, to use the iPhone 4's front-facing camera on yourself.

Shooting video

To switch between stills and video shooting, toggle the little switch, bottom-right. Though you can't enable the zoom feature for video, from here on the process is pretty much the same as for stills, but with a glowing red light on the stop/start button so that you know when you are shooting.

Tip: When shooting video in low light, switch the ⚡ button to On for continuous illumination.

Removing the camera

In the US, several government and military organizations, and even a few commercial businesses, refuse to let employees bring camera phones onto their premises due, supposedly, to espionage risks. If that applies in your workplace, you may want to have the camera removed. A company called iResQ will do the job for $100, including overnight shipping in both directions.

iResQ iresq.com/iphone

The Camera Roll

The images and videos you take are saved together in the so-called Camera Roll, which can be found, when using the camera, by tapping the preview of your last shot, bottom-left. Alternatively, look within the Photos app. Videos appear here with a ▶ icon in their centres. Tapping the screen reveals additional controls and a scrubber bar for moving back and forth through the footage.

Photographing contacts

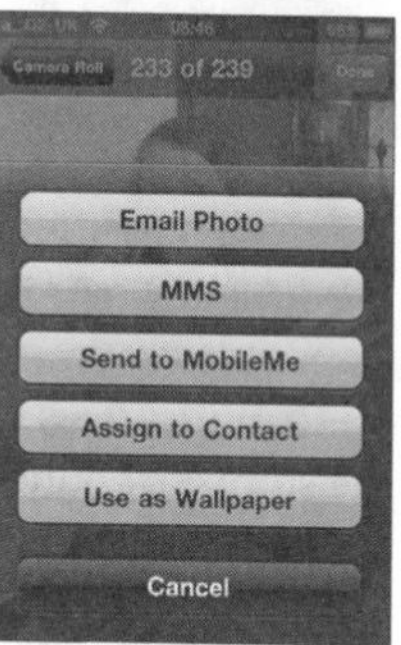

One thing you can do with the camera is shoot pictures of friends and family and then assign them to relevant entries in the Contacts lists. Annoyingly, though, you have to create a contact entry first (see p.93) and add the image there; you can't create a new contact directly from a picture.

Video editing

You can snip the ends off video clips straight from the Camera Roll preview (pictured above); simple drag in the two end points on the scrubber bar and then tap the Trim button, top-right.

For more sophisticated editing, transitions, themes, music, and the like, you will need a dedicated app such as Apple's own iMovie. Though some of the themes might seem a little cheesy, the timeline and transition tools work very well on the small screen.

Apps for your camera

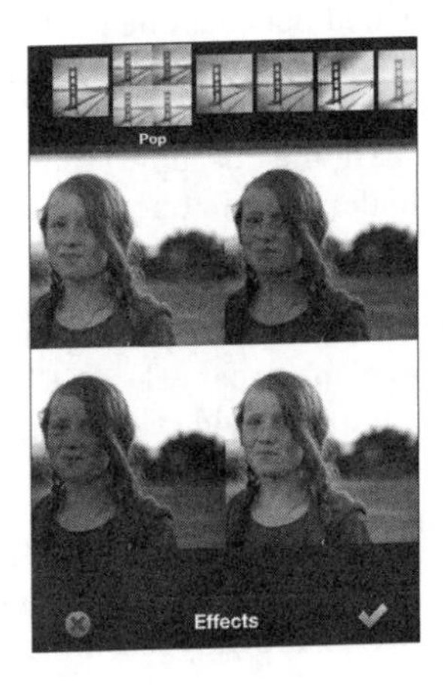

• **CameraBag** A really nice collection of filters and visual effects to enhance your iPhone photography.

• **Camera+** This is an excellent, fully-featured alternative to the iPhone's built-in Camera app. It's special image stabilizer controls work really well to stop the wobble.

• **FatBooth** Pile on the pounds with this little app that shows what you would look like if you were more interested in pies than iPhones. The same developers also make AgingBooth.

• **Hipstamatic** Relive those analog days of oversaturated, grain-heavy snaps with this very popular app.

• **Pano** Excellent app for creating stitched panoramas from multiple shots.

• **PS Mobile** (Pictured) This is Adobe's iPhone-friendly version of their popular Photoshop desktop application. It's really easy to use and has some nice effects as well as a useful crop tool.

• **Tiltshift** Create interesting focus effects within existing images.

• **Video Camera (for iPhone 2G and 3G)** That's right, a video camera app for pre-3GS iPhones that don't support video recording straight out of the box.

Putting existing pics on an iPhone

The iPhone can be loaded up with images from your computer. iTunes moves them across, in the process creating copies that are optimized for the phone's screen, thereby minimizing the disk space they occupy. iTunes can move images from an individual folder or from one of three supported photo-management tools:

• **iPhoto (Mac)** apple.com/iphoto
Part of the iLife package, which is free with all new Macs (and available separately for $79). Version 4.0.3 or later will sync photos and videos with an iPhone, but version 6 is better, enabling you to view pictures according to the faces in them.

• **Aperture (Mac)** apple.com/aperture
Apple's professional photo suite can do the job, but is pricey ($199) and only recommended to the dedicated photographer.

• **Photoshop Elements (PC)** adobe.com/photoshopelements
Like the above, but with far more editing tools and a $90/£60 price tag. You'll need version 3.0 or later.

To get started, connect your iPhone and look for the options under the Photos tab in iTunes. Check the sync box then choose your application or folder.

How to stop iPhoto popping up

If you sync your iPhone with iPhoto, when you connect the phone to your Mac, iPhoto may automatically launch and offer to import recent snaps taken with your camera phone. (If your Camera Roll is empty, this won't happen.) The way around this is to open iPhoto > Preferences and from the General tab, choose "Connecting camera opens: no application" from the dropdown of options.

Viewing images on the iPhone

Once your images are on the phone, photo navigation is very straightforward. Tap Photos and choose an album – or tap Photo Library to see the images in all albums. You can also choose to tap Places to see all photos that have associated GPS location data represented as pins on a map.

> **Tip:** If you have synced photos from iPhoto based on its Faces feature recognition tool, tap Faces to view the list. Unfortunately, the Photos app only displays synced Faces sets and cannot add to them as you take new photos.

When viewing a set of images:

• **Share** Tap , highlight the images you want to send and then tap Share to add them to an email or MMS message

• **Copy** Tap , highlight the images you want and then tap Copy.

When viewing individual images:

•**"Flick" left and right** to move to the previous or next photo.

• **Zoom in and out** Double-tap or "stretch" and "pinch" with two fingers.

• **Rotate the iPhone** to see the picture in landscape mode.

• **Hide or reveal the controls** Tap once anywhere on the image.

Online image posting from the iPhone

With a MobileMe account (see p.27) it's very easy to post images directly from the iPhone. First, however, you have to enable the feature in iPhoto for a specific published album; from then on, simply tap the button when viewing a photo on the iPhone and select Send to MobileMe. For the full story, head to the MobileMe website and browse the Learning Center.

In the App Store, there are hundreds of apps for photo sharing and uploading waiting to be discovered. Flickr, the world's most popular photo-sharing site, has an excellent free app that account-holders can use on an iPhone. Pixelpipe, meanwhile, allows you to upload pictures to Flickr, Picasa, Facebook, YouTube and dozens more sites, all from one place.

It's also worth noting that the Facebook app (see p.186) and nearly all Twitter-interfacing apps also allow you to post using your camera.

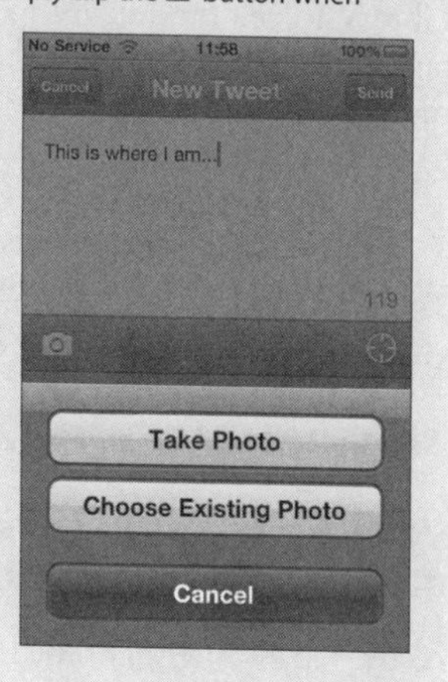

- **More options** Tap to assign it to a contact, send as an MMS or email, use it as wallpaper, or upload it to MobileMe (see box above).

- **Start a slideshow** Tap ▶. By default, the phone will show each photo for three seconds, but you can change this by tapping Settings > Photos. The same screen lets you add Star Wars-esque transitions, and turn on Shuffle (random order) and Repeat (so that the slideshow plays around and around until you beg it to stop). You can also connect to a TV (see p.169) to play your slideshows on a big screen.

18

Maps

Search and directions

The Maps app on the iPhone (tap the icon on the Home Screen to get started) takes you into the world of Google Maps, where you can quickly find locations, get directions and view satellite photos. You can zoom and scroll around the maps in the same way you would with webpages in Safari, the difference being that, because you're using your own digits to drag and pinch, the whole experience feels much more natural.

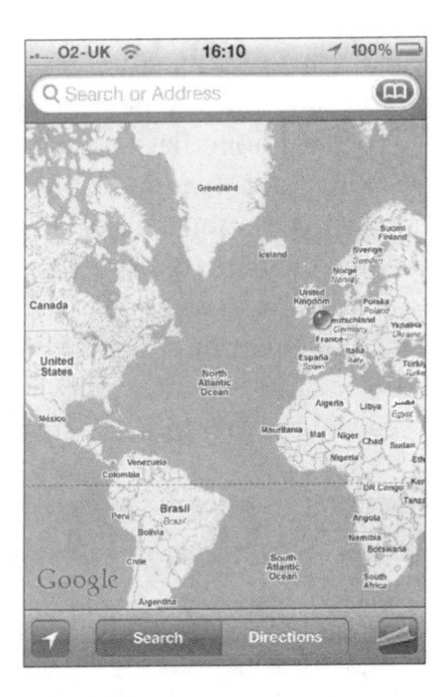

Tip: To zoom in, either double-tap one finger or spread two apart; to zoom out again, either pinch or tap once with two digits.

Searching for yourself

All recent iPhones can accurately determine your current location by using a combination of data from their GPS (Global Positioning System) chips, a connected cellular network and also from a connected Wi-Fi network. The first generation iPhone, on the other hand, has no GPS chip and so has to rely on the cellular network and Wi-Fi. To pinpoint your location, open Maps and tap the ➤ button at the bottom of the screen. If GPS isn't available, or you're using a first-generation iPhone, you'll get a less accurate result.

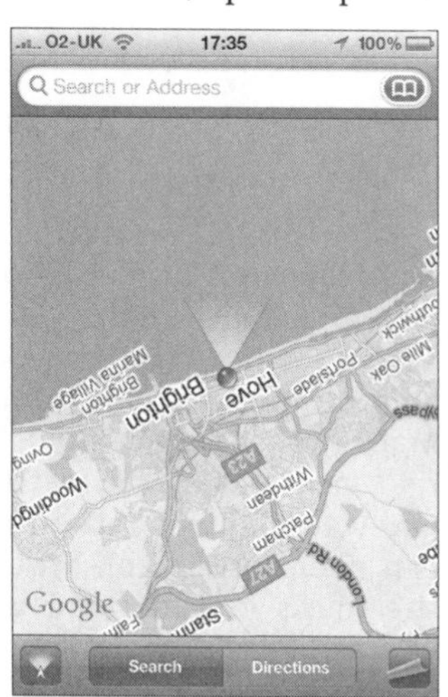

Which way am I facing?

If you have an iPhone 4 or 3GS, hold the iPhone flat and tap the ➤ button for a second time; the app switches mode, allowing you to view the map in relation to the way you are facing. The tighter the angle coming out of your blue location marker, the more accurate the direction reading.

Location Services settings

Whenever the app you are currently using is hooked into Location Services (via either GPS, Wi-Fi or the cellular network) a purple ➤ icon appears in the status bar at the top of the iPhone's screen. Look within Settings > General > Location Services to turn your iPhone's location functionality on and off for either specific apps or everything. This can be a really useful means of saving battery power. You can also see here which apps have used the iPhone's Location Services in the last 24 hours as they display a purple ➤ icon.

Searching for a location

Tap the Search box and type a city, town or region, place of interest, or a ZIP or postcode. (As you type, the iPhone will try to predict the location based on previous map searches and address entries in your Contacts list.) You can also try to find a business in the area you are viewing by entering either the name of the business or something more general – such as "camera", "hotel", or "pub". Note, however, that the results, which are pulled from Google Local, won't be anything like comprehensive.

If a multitude of pins appears, tap on each of the pins in turn to see their name. Tap a pin's blue ❯ icon for further options, such as adding the location as a bookmark or contact address, emailing a link to the location or getting directions to or from that location.

> **Tip:** In some areas you will also see a little red icon on a pin's name bubble. Tap this to scroll around in Google's Street View. When you're finished, tap the inset map preview to finish.

Dropping pins

You can also drop a pin manually at any time by tapping the page-curl button, bottom-right, and then Drop Pin. This can be a handy way in which to keep your bearings when sliding around a map. To change the position of a pin, tap and hold it and then drag. To delete the pin, add it as a bookmark, add it to a contact, email the location or get directions, tap the pin and then the ➋ icon. Your lists of bookmarks and recently viewed locations can be viewed by tapping the 📖 icon in the right of the search field.

Satellite and Terrain views

Tapping the page curl button reveals options to view Satellite images (they're not live, unfortunately … maybe one day), a Hybrid view that adds roads and labels from the standard view to a satellite image, and also a List view, which can be a handy way to quickly take in all the pins that were generated by a search.

Directions

To view directions between two locations, tap Directions at the bottom of the screen and enter start and end points, either by typing search terms or by tapping 📖 to browse for bookmarks, addresses from your Contacts and recently viewed items. When you are ready, tap Route. Use the three icons at the top to choose between directions for driving, walking or taking the bus.

Once a journey is displaying, you can go through it one step at

> **Tip:** Where a search term that yielded multiple results has been used as an end point, tap the ☰ icon to quickly switch between the different pins, and in turn, refresh the route.

a time by tapping Start and using the arrow buttons to jump forward and back one stage. Alternatively, tap the page-curl button and then List to view all the stages as a series of text instructions. Tap any entry in the list to see a map of that part of the journey.

> **Tip: For your return journey, reverse the directions given by your iPhone by tapping the ⮃ button between the start and end point fields.**

Traffic conditions

In areas where the service is available, your route will display colour-coded information about traffic conditions. The approximate driving time at the top of the iPhone screen will change to take account of the expected traffic and the roads will change colour:

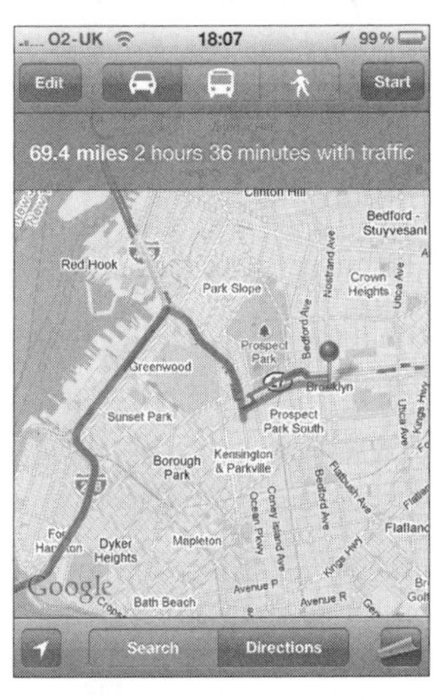

- **Grey** No data currently available.
- **Red** Traffic moving at less than 25 miles per hour.
- **Yellow** Traffic moving at 25–50 miles per hour.
- **Green** Traffic moving at more than 50 miles per hour.

If you don't see any change in colour, you may need to zoom out a little. This action will also automatically refresh the traffic speed data. To hide the traffic information (perhaps you like surprises), tap the map curl and toggle the Traffic button off.

Maps, apps and navigation

The App Store has its own category dedicated to Navigation, where you'll find everything from plane and ship location tools (Plane Finder and Ship Finder) to multi-functional compass applications (Harbor Compass Pro; a great alternative to the iPhone's built-in Compass app). Also search "maps" to unearth some interesting cartographic relics (Historic Earth) as well as some very impressive reference tools (National Geographic World Atlas). And a few more…

- **Google Earth** An essential free download that gives you unfettered access to the Globe.

- **Transit Maps** Use this app to locate, download and store transit maps from around the world that you can then use offline. A good alternative for London is Tube Deluxe.

- **iTopoMaps** Beautifully rendered maps for hiking and climbing in the US.

- **iOS Maps** A similar app to the above, but for the UK, courtesy of Ordnance Survey.

- **London A-Z** Brilliant set of London street maps. It's all stored offline so you don't even need a web connection to use it. There are equivalent offline maps available in the App Store for most major cities – essential if you know you are going abroad and don't want to rack up roaming charges by using Google Maps.

- **Wikihood** Location-based access to Wikipedia … great for finding out about buildings around you.

19

Reading

eBooks, newsfeeds and more

Despite its small screen, the iPhone makes a surprisingly good eBook reader – especially since the launch of Apple's iBook app and iBookstore, which makes thousands of books available at the touch of a button.

At the time of writing, iBooks is not one of the default apps on the iPhone. It's free, but you will have to visit the App Store to download it.

Once launched it displays a pretty bookshelf (your Library, where you can browse all your virtual tomes). To toggle to the iBookstore (which looks very similar to the iPhone's iTunes Store), tap the button in the top-right corner of the app.

Using the iBookstore

There really aren't any surprises here. You browse books and purchase them in just the same way that you do music, movies and TV shows in the iTunes Store. You can even use the same log-in details. Start exploring and you'll quickly get the hang of it.

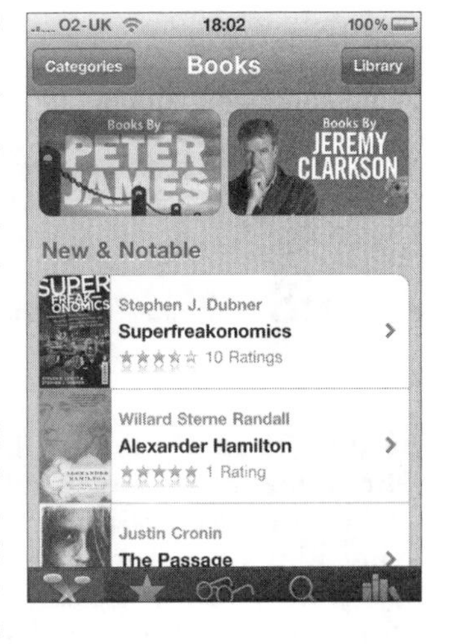

Sample chapters

In most cases you can tap Get Sample to download a free excerpt to whet your appetite. You can download the full text at any time while reading (assuming you are connected to the Internet) by tapping the Buy button at the top of any page in the sample.

Tip: **You can toggle between the "bookshelf" and "list" views of your library using the two buttons near the top.**

Reading with iBooks

To open a book (as opposed to a PDF, see p.218), simply tap the Books tab within your iBooks Library and then choose the title you want to read. To turn a page, either drag the bottom corner, or tap to the left or right of the text near the edge of the iPhone's screen.

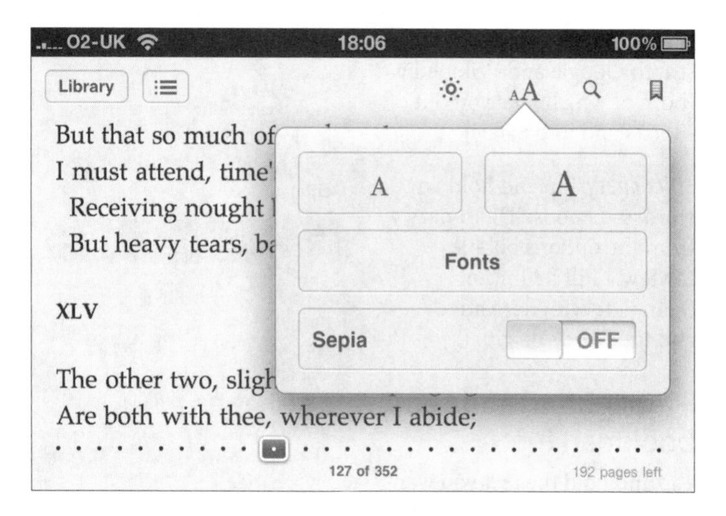

A single tap anywhere on the body of the page will reveal or hide further options at the top and bottom, including brightness, a bookmark, text options and text search. This tap also reveals a slider at the bottom of the page to help you quickly jump to another part of the book.

- **Contents page** Tap the ▤ button to view the Contents page of the title you are reading; once there, tap Resume to return to the point where you were reading.
- **Search** Tap and hold any word and choose Search from the options bubble; this displays a tappable list of other places where

> **Tip: Look within Settings > iBooks to determine whether tapping on the left-hand edge takes you to the next page or the previous page. Here, you can also turn text "justification" on and off … try it both ways to see what you prefer.**

the word occurs and links to search Google and Wikipedia (both of which take you out of iBooks and into Safari).

- **Dictionary** Tap and hold any word and choose Dictionary from the options bubble to view a full definition (complete with derivatives and the word's origins).

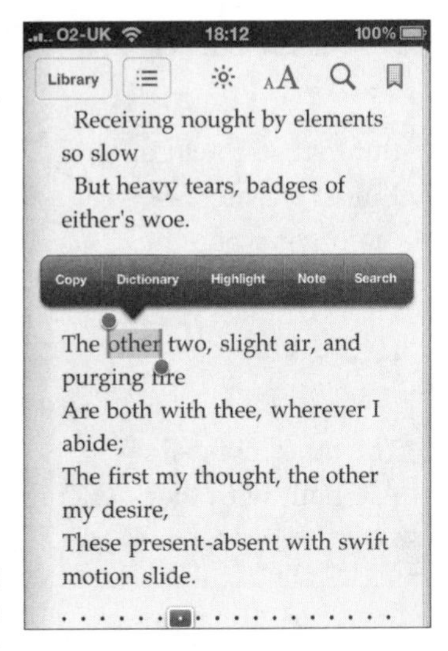

Highlighting and bookmarking

Tap and hold the text you want to highlight and then drag the blur anchor points to adjust the size of the selection. When you are ready, tap Highlight. You can then tap the highlighted text to either change the colour, add a note or remove the highlight.

To view a list of all the highlighted sections within a given book, tap the ▤ button at the top of the page, and then tap Bookmarks. From here you can also delete specific bookmarks from the list by swiping across them and then tapping Delete.

Reading PDFs

When viewing a PDF sent as an attachment in an email (see p.193) or online within Safari (see p.184), you should see an "Open in iBooks" button which will add the PDF to your iBooks library. Once this is done, the file will become available on the PDF section of your bookshelf – even when you are no longer connected to the Internet.

The reader controls for PDFs are very similar to those for books, but without the specific text options. If you don't get on with iBooks as a PDF reader, try an alternative app such as the excellent GoodReader.

Deleting books and PDFs

To delete books from your iPhone's Library, tap Edit and then the ⊗ icon on the title you want to ditch. Assuming it's been synced across already to your Mac or PC (see below), the title will still be available to sync back onto the iPhone later, should you wish to do so.

> **Tip: If you are looking for a way to read ePub files on your Mac or PC, try the free-to-use Adobe Digital Editions (adobe.com/products/digitaleditions).**

Syncing with iTunes

To sync your books with iTunes, connect your iPhone to your Mac or PC, select the Books tab, and choose from the options. Once everything is set up, new iBookstore downloads will be synced back to iTunes every time you connect, along with any PDFs that you have added to your library.

> **Tip: If no Books listing appears under Library in the iTunes sidebar on your computer, open iTunes Preferences and check the Books box on the General pane.**

ePub books from other stores

iBooks can also display books in the ePub format from sources other than the iBookstore (assuming they don't have any special

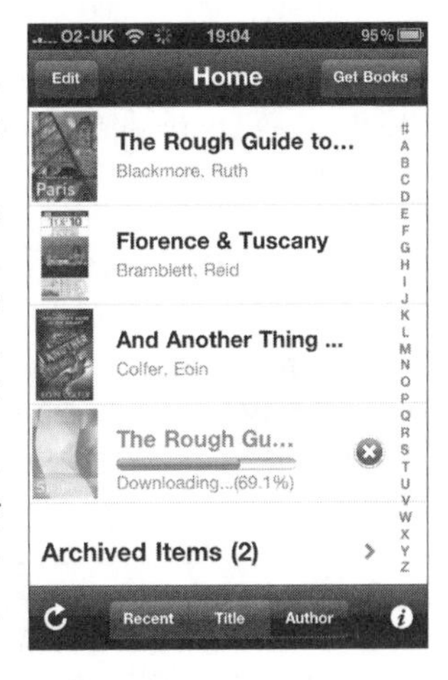

DRM copy protection built in). First, you need to get them into iTunes on your computer. Highlight the Books listing in the iTunes sidebar and then drag and drop the files into the main iTunes window, where they will appear alongside your iBookstore purchases. You can then connect your iPhone and sync them across in the normal way.

This won't work for Sony Reader Store titles (which employ a form of DRM), but there are plenty of other ePub sources that you can turn to, including:

Google Books books.google.com
ePubBooks epubbooks.com

Whether you are an author yourself, a student, or someone who simply wants to view and organize text documents within the iBooks app, there are loads of benefits to being able to create your own ePub files. To give it a go, try:

Storyist storyist.com
eCub juliansmart.com/ecub

Other readers

Using iBooks is not the only way to read eBooks on the iPhone. And when it comes to specialist graphic titles, such as comics and mangas, you are far better off turning to the App Store than the iBookstore to get your fix.

eBook reading apps

• **Kindle** For those who want to purchase their reading material from Amazon.

• **Stanza** This reader/store app offers thousands of titles and you can also add your own files using the Stanza Desktop application.

Comics and graphic novels

• **Marvel** Access to an essential store of classic superhero comics. Download and read either page by page or frame by frame.

• **iVerse Comics** A really easy-to-use store and comic reader. There are both paid-for and free comics to be found here covering all genres and ages of reader.

• **iMangaX** More than a thousand manga to browse and read. All the content is stored online, so an Internet connection is essential.

News and RSS feeds

The iPhone is ideally suited to delivering your daily dose of current affairs, or any other kind of news for that matter, wherever you want it, and in a convenient format. News can be found via Safari, of course, but news is best read via either an aggregating RSS tool or a dedicated app from one of the main news providers.

RSS

RSS – Really Simple Syndication – allows you to view "feeds" or "newsfeeds" from blogs, news services and other websites. Each feed consists of headlines and summaries of new or updated articles. Right now, arguably the best web-based aggregator is Google Reader, which can be accessed via Safari (reader.google.com). To get started, simply sign in using the same credentials you use for other Google services and start adding feed.

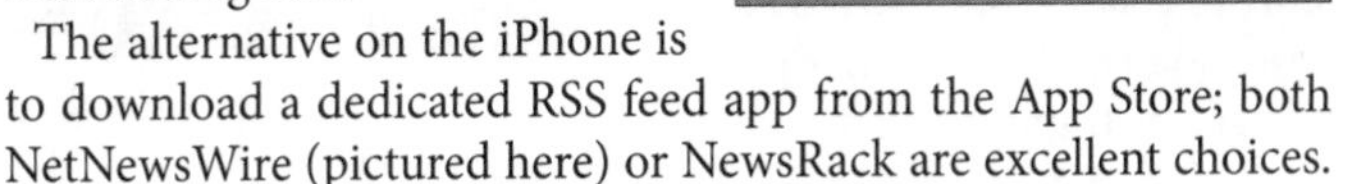

The alternative on the iPhone is to download a dedicated RSS feed app from the App Store; both NetNewsWire (pictured here) or NewsRack are excellent choices.

Also worth investigating is the free Instapaper service, which lets you save webpages and articles to read later. The Instapaper app has a really easy-on-the-eye stripped-back feel; many RSS services (including NetNewsWire) integrate with it, as does Twitter, so you can have all your reading matter and stories with you, and available offline (assuming you remember to sync your app and Instapaper account before you disconnect from the Internet).

News apps

Most of the major news and magazine publishers are battling to define their place on the iPhone and the wider digital market-place. Some have opted to make their applications free, but with advertising support; some others are looking to get users to pay a subscription; while others will charge a one-time fee for users to download their app.

The real question here is not whether you want to pay for your news (there are still a thousand places online where the latest stories can be harvested for free), but more whether you are prepared to pay for a handy, usable app and a particular editorial voice or standard.

The best news apps include:

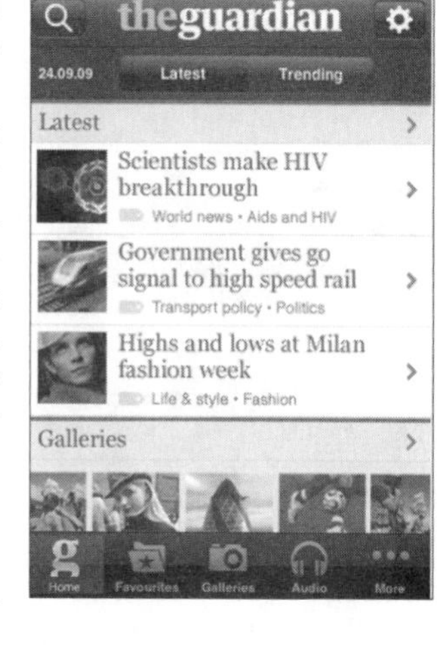

- **The Guardian** Great-looking UK news app (pictured) with loads of photography and a customizable download feature, so that you only get the news categories you want.

- **The Wall Street Journal** Register for free to see the latest world and market news.

- **USA Today** A nice interface with an easy-to-navigate layout. It downloads new content very quickly.

- **The Financial Times** If you are prepared to pay the monthly subscription fee, this is, as you might expect, the best app around for stocks and business news. If you are not prepared to register, then you only get to look at a handful of articles each month.

- **Zinio** A one-stop shop for magazines. There are hundreds on offer, plenty of free samples, and you can purchase single issues and annual subscriptions. There's a handy calendar view that shows you which issues you'll be getting to read each month.

20

More tools

Apps for everything...

Like most modern phones, the iPhone comes with a small menagerie of extra tools such as alarms, calendars, a calculator, a weather app and a notes tool. But thanks to the App Store, the iPhone apps that come built in are really just the tip of a hulking great iceberg. This chapter glances briefly at a few of the most useful tools offered by both Apple and third-party developers.

Apps that keep you organized

Calendar

The iPhone Calendars app (note that the Home Screen icon always shows the right date) can be synchronized with your home computer using iCal or Entourage on a Mac, or Outlook on a PC, and is managed within iTunes. Select your iPhone's icon and look for the options on the Info tab.

> **Tip:** You can set an alert to remind you of an impending event either as it happens or a certain number of minutes, hours or days beforehand. If you'd like to have these alerts presented visually, instead of via a sound, tap Settings > Sounds and turn off the Calendar Alerts.

Alternatively, you can sync calendars over the airwaves via a server using Exchange, MobileMe or Google (see p.187). These are set-up directly from the iPhone within Settings > Mail, Contacts, Calendars. Once syncing is set up, calendar data is merged between your computer and phone, so deletions, additions or changes made in either place will be reflected in the other next time you connect (when syncing via iTunes) or almost instantly, when you "push" your changes via a server (see p.190).

You can view your schedule by month or day, or in a list – though annoyingly not by week. To add a new event, tap **+**, enter whatever data you like, and tap Save. To edit or delete an existing event, tap the relevant entry and use the Edit button or Trash icon.

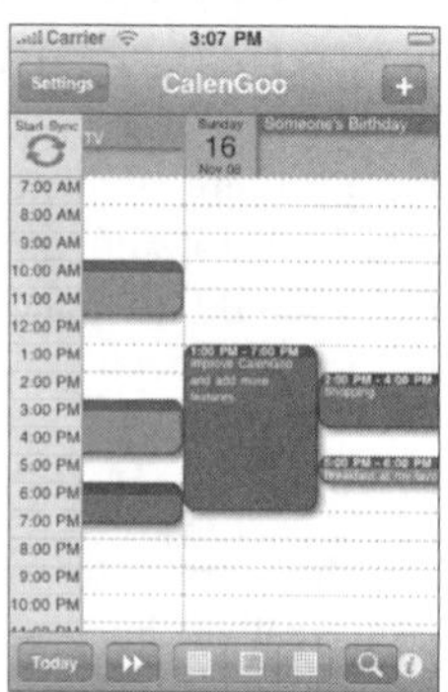

Calengoo

If you use Google Calendars, then this app is an excellent alternative to the built-in Calendars tool. It has a really nice look and feel, and even supplies the essential "week" view missing from the Apple app.

Clock

To quickly check the time on your iPhone, look to the digital clock which is almost always present at the top of the screen. The Apple Clock app, meanwhile, offers a World Clock (tap **+** or Edit to add and remove locations), an Alarm (use the **+** button to add as many alarms as you like) and a self-explanatory Stopwatch and Timer.

> **Tip:** **Even if you have your iPhone in silent mode, the alarm will still sound when it is due to go off – which is one less thing to worry about when setting your alarm at bedtime, but could, potentially, cause embarrassment in the cinema or a meeting.**

GeeTasks

A slick little app for Google Task syncing. You even get a badge on the Home Screen icon that displays how many tasks you have yet to complete.

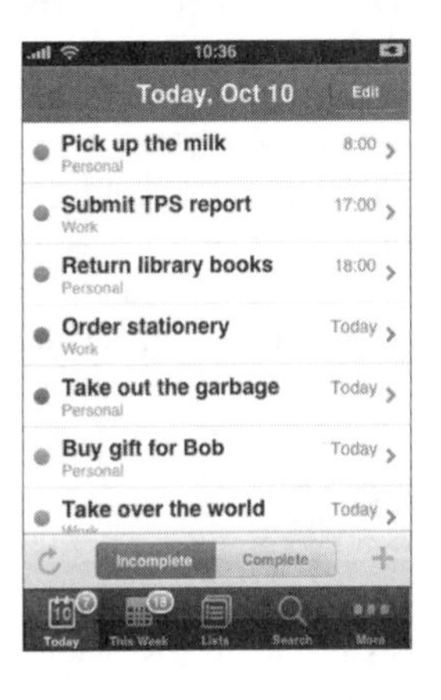

Remember The Milk

Pictured to the right, this app is arguably the best task-organizing sync service out there right now. Unfortunately you have to sign up for a fee-paying Pro account to make use of the free iPhone app.

Toodledo – To Do List

Another service that syncs your tasks to the cloud; a nice interface and excellent integrated noting tools.

Note-taking apps

Notes

Notes is Apple's simple application for jotting down thoughts on the go. Tap **+** to bring up a blank page and you're ready to type. When you're finished, tap Done or use the envelope icon to email the note.

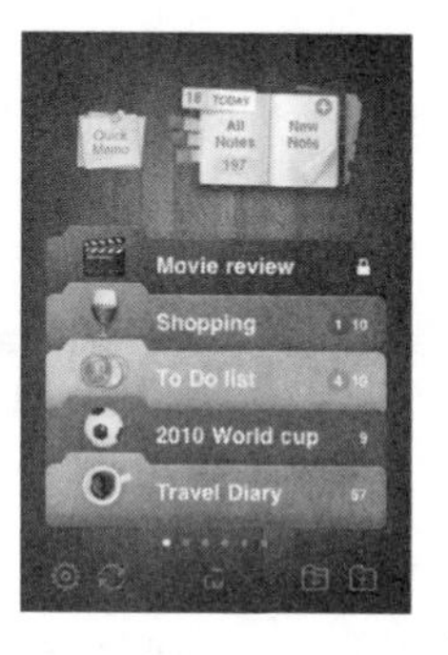

Awesome Note

A really stylish notes and to-dos tool with colour-coded folders and lists. (Pictured right.)

Evernote

Evernote is among the best note-taking tools and services available. The iPhone app can create notes from text, images and audio and these notes are then synced between the Evernote server and whatever desktop or mobile versions of Evernote you are running.

Voice Memos

For voice notes, don't forget that the iPhone comes with an excellent memo recorder built in. Recorded memos can be sent via email or MMS from the iPhone, and also synced back automatically to a special Voice Memos playlist in iTunes when you connect.

> **Tip: You can use Calendar instead of a notes app to jot down thoughts on the move. Just add an event and use the Notes field. You can set a date and time when you think the jottings will be useful, and even set an alert.**

Writing apps

Doc²

The formatting tools of this word processor app (pictured) are really impressive, offering everything from bold, italic and underline to bullets, indents and table construction. Its best feature is the fact that it works with Google Docs to help you share your files online.

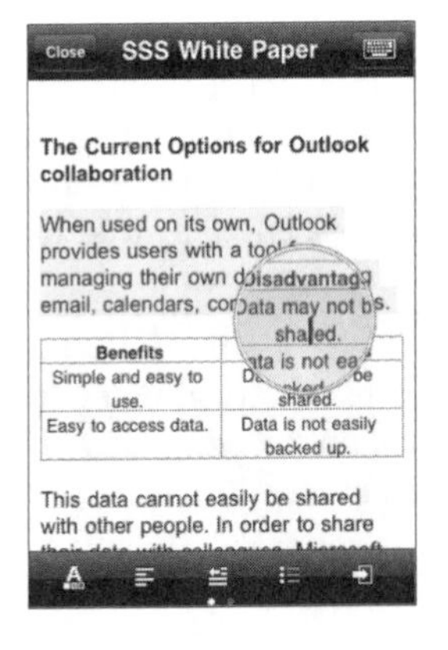

WordPress

For the WordPress bloggers out there, the iPhone app is a must, both well designed and easy to use. As well as creating and editing your posts, you can also use the tool to moderate comments.

WriteRoom

This app offers a distraction-free writing environment where you can get on with your writing without toolbars and formatting worries. It works in a white-on-black mode that makes it easy on the eye for extended periods of writing.

Pocket CV

This app makes sure you always know where your résumé is, and more importantly, keep it up to date. Its layout features are impressive, meaning you can email off a fully fledged, PDF CV at the drop of a hat from anywhere.

Money apps

Stocks

The built-in Stocks app lets you view current values for any listed company. Well, not quite current – the prices are refreshed whenever you open the application, but are typically still about 20 minutes out of date.

XE Currency

XE Currency (pictured), is the only currency converter tool you need. It really comes into its own when you're abroad and need to quickly check whether or not something is a bargain.

Weather apps

Weather

The iPhone's built-in Weather application does pretty much what you would expect. You can save favourite locations, view current conditions along with a six-day forecast with temperature highs and lows.

Met Office Weather Application

For the UK, this Met Office alternative is a must, largely because it includes radar and satellite imagery. Also, the five-day forecast tends to be more accurate than the one on the Apple app (which pulls its data from Yahoo!).

Number apps

Calculator

Paying no small tribute to the classic designs of German functionalist Dieter Rams, the look of the iPhone's built-in calculator app can be traced back to a calculator that Braun (for whom Rams worked) produced for Apple in the mid-1980s. Turn the phone through ninety degrees to reveal a fully featured scientific mode.

PCalc Lite

Though you can, arguably, get everything you need from Apple's Calculator app, PCalc Lite is free, looks nice, boasts some scientific functions and can also handle unit conversions.

Quick Graph

This free app can be used to produce both 2D and 3D graphs from complex formulas and equations.

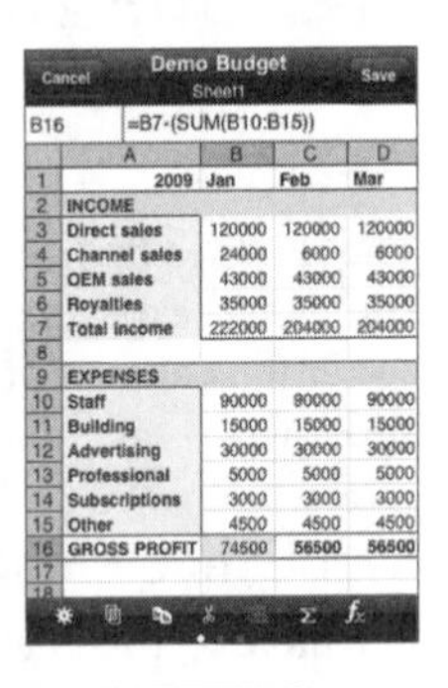

	A	B	C	D
1	2009	Jan	Feb	Mar
2	INCOME			
3	Direct sales	120000	120000	120000
4	Channel sales	24000	6000	6000
5	OEM sales	43000	43000	43000
6	Royalties	35000	35000	35000
7	Total income	222000	204000	204000
8				
9	EXPENSES			
10	Staff	90000	90000	90000
11	Building	15000	15000	15000
12	Advertising	30000	30000	30000
13	Professional	5000	5000	5000
14	Subscriptions	3000	3000	3000
15	Other	4500	4500	4500
16	GROSS PROFIT	74500	56500	56500
17				

Sheet²

This app crafts fairly impressive spreadsheets and can both edit and create Microsoft Excel docs. It also keys in with Google Docs for sharing your files online.

Tip: If you like the sound of both Doc² and Sheet², check out the Office² app, which combines both at a cheaper price.

Productivity apps

Jotnot Scanner Pro

A useful scanning tool. Use the iPhone's camera to take a photo of some text and then store the resulting file or send it to colleagues, friends or family. Shadows are automatically removed.

Air Display

Turns your iPhone into an extra screen for your Mac. Use it, for example, to display a small-window application such as iChat or Skype while you use your larger Mac screen for other purposes.

File storage apps

MobileMe iDisk

A free, simple tool for subscribers to Apple's MobileMe service to view the contents of their iDisk.

Dropbox

Dropbox is probably the best online file-storage system, making it a cinch to sync files and folders between multiple computers. Best of all it's free – including this iPhone app which allows you to view the contents of your DropBox folder from your phone.

Air Sharing HD

This excellent app gives onboard file storage, Wi-Fi file sharing with desktop machines and access to networked printers.

And finally...

Gourmet Egg Timer

For the perfect boiled egg, this app takes into account altitude, egg size and how soft you like your yolk.

Methodology – Creative Inspiration Flashcards

Modelled on the "Oblique Strategies" deck of cards created and originally published back in 1975 by Brian Eno and Peter Schmidt, this app presents a set of cards that, in their own words, "prompt you to think differently". They can be really effective (and at other times simply amusing) for those moments when inspiration just doesn't come.

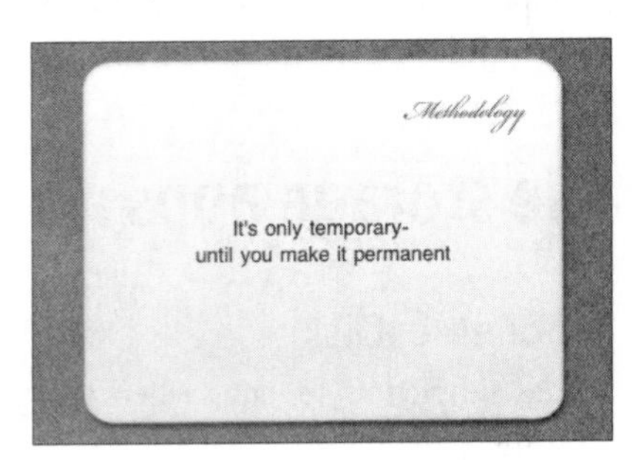

Bloom

Speaking of Brian Eno, for a bit of ambient music generation, download his rather lovely Bloom app.

iPhonology

21

Accessories

plug and play

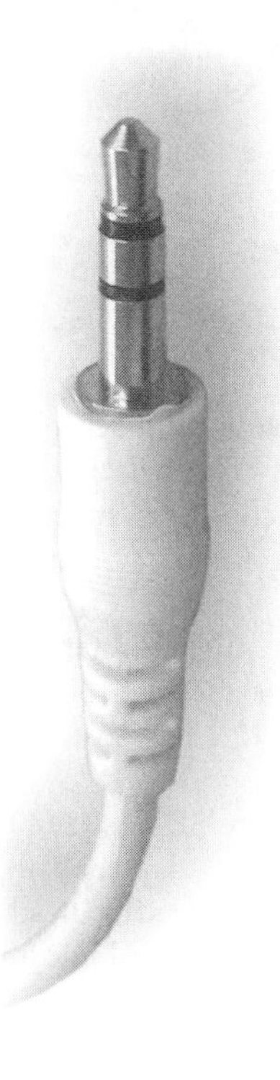

There are scores of iPhone accessories available, from FM transmitters to travel speakers. The following pages show some of the most useful and desirable add-ons out there, but new ones come out all the time, so keep an eye on iPhone news sites (see p.246) and the Apple Store's iPhone department. When it comes to purchasing, some accessories can be bought on the high street, but for the best selection and prices look online. Compare the offerings of the Apple Store, Amazon, eBay and others, or go straight to the manufacturers, some of which sell direct. Before buying any accessory, make sure it is definitely compatible with your iPhone model. Some, but certainly not all, accessories designed for the iPod will work with an iPhone.

Wired headsets

Models include: **Rivet Mobile Stereo Headset; V-MODA Vibe Duo; Shure SE210**
Cost (approx): **$30–100**

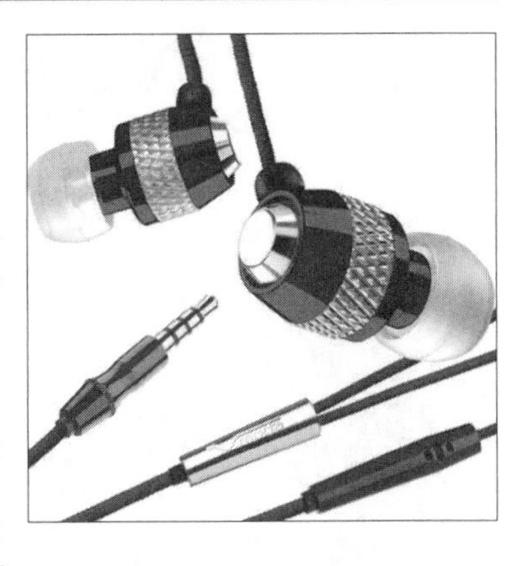

The stereo headset that comes bundled with the iPhone is functional enough, but not ideal. Even if the audio quality was better, it still wouldn't be especially comfortable to wear. Many people can't even get the earphones to stay in place.

Thankfully, there are numerous alternatives out there which are comfortable, "noise isolating" (that is, they snugly fill your ear thereby blocking out ambient noise) and have a minijack plug svelte enough for the iPhone's recessed port. The best budget option is Rivet's Stereo Headset ($30), which offers superior sound, lanyard-style cords, a decent mic and a choice of interchangeable earbuds.

If you're serious about sound quality, you might prefer the V-MODA Vibe Duo (pictured top-right). These cost $100 but offer crisper, clearer sound, perhaps thanks to their so-called BLISS noise-cancelling system.

The best we have tested, however, are the Shure SE210s ($180) in combination with a Shure MPA, Mobile Phone Adapter ($50), which gives you an unbeatable pair of sound-isolating headphones in conjunction with an iPhone-compatible lanyard mic for making and answering calls.

Bluetooth headsets

Models include: **Aliph Jawbone**
Cost (approx): **$50–130**

Bluetooth headsets let you receive calls without getting out your phone or wearing a wired headset. They sit in or around your ear, respond to voice commands and communicate with your phone via radio waves. The main problem is that they can make you appear to be talking to yourself – so expect some strange looks in the street.

Apple's own offering (pictured top-right) was released at the same time as the iPhone but quietly discontinued in 2009 – presumably because of low sales. There are many other brands still available, though. One slick-looking but pricey option is the Aliph Jawbone ($100, pictured below). Its over-the-ear clip is very comfortable and it boasts some clever built-in software that adjusts the volume to account for the noise of your environment.

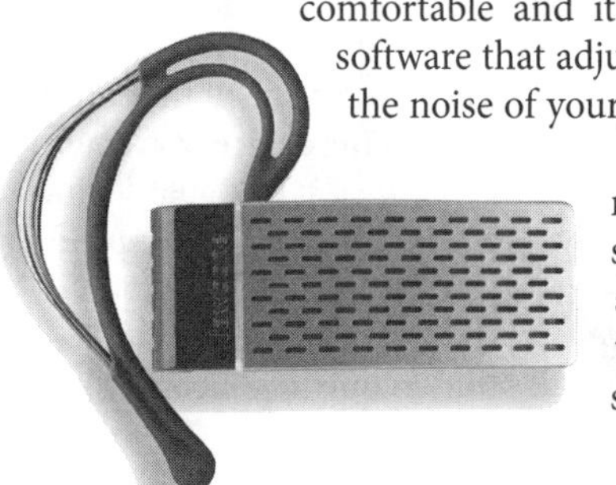

For those on a tighter budget, most other Bluetooth headsets should work fine. They start at around $50, though the very cheapest ones tend to suffer from poor battery life.

Cases and screen protectors

Models include: **Pacific Rim InvisiShield & iShield; iGone Silicone Skin; Belkin Sports Armband; Incipio Bikini Case; Miniot iWood**
Cost (approx): **$15–50**

The iPhone is more resistant to scratches and other day-to-day wear than the iPod and many other smartphones. But it's not invulnerable, so you might want to protect yours with some kind of case.

One problem with cases is that they're unlikely to do much for the look of the device, so you might opt instead for an invisible screen protector such as the iPhone InvisiShield, pictured right ($15 for a pack of two, from pacrimtechnologies.com).

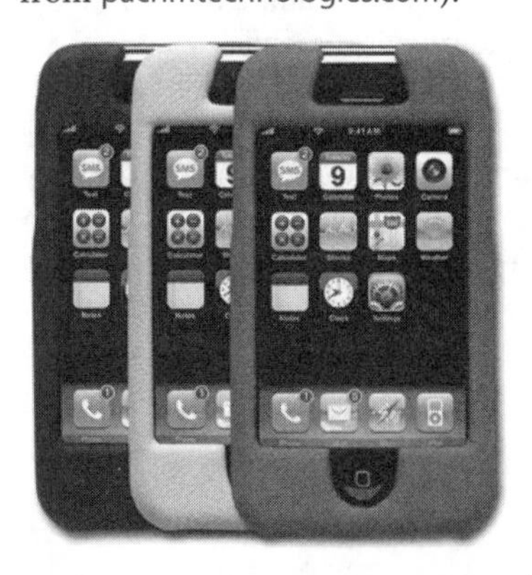

For protection of the entire body of the device, you could combine a screenguard with a slip case such as Apple's own Bumpers for iPhone 4. Or, for older iPhone models, the colourful Silicone Skin case from iGone ($13 from igonemobile.com) or Pacific Rim's iShield, pictured left, which comes with a synthetic, leather-ish finish.

If you want to use your iPhone when jogging, or in the gym, look for something with an armband (such as Belkin's Sports Armband, pictured right) or a belt clip.

As with iPods, it didn't take long for more unusual iPhone cases to emerge. These range from the Kharki Bikini Case, pictured bottom left ($20 from myincipio.com), to the hand-carved Miniot iWood, picture bottom right (price yet to be confirmed, from miniot.com).

Some users have pointed out that, since the iPhone gets quite hot, wrapping it in anything that stops the dissipation of heat might cause damage to its internal components. But that seems unlikely – and certainly offset by the reduced risk of damage by dropping.

iPhone speaker units

Models include: Bose SoundDock; Soundmatters foxL2; Logic3 i-Rotate
Cost (approx): $50–350

An alternative to connecting your iPhone to a hi-fi or a set of powered speakers is to use a self-contained iPhone/iPod speaker system. These are easy to move from room to room (or indeed, from place to place when travelling) and they are space efficient, too. There are literally hundreds of models available, many of which can run on batteries for extra portability.

There are audiophile options in all shapes and sizes, from the tiny Soundmatters foxL2, which can be held in one hand, to mid-size systems such as the popular Bose SoundDock (pictured above). At $300, the latter isn't cheap, but the sound is exceptionally clear and punchy for the size. The Logic3 i-Rotate, meanwhile, can hold the iPhone in either portrait or landscape mode, enabling videos to be viewed full-screen.

For a while, Apple made its own speaker unit, the chunky iPod Hi-Fi (left), which (given enough batteries to power a small village) could be taken out and about ghettoblaster-style. It was discontinued in 2007, though can still be found on eBay.

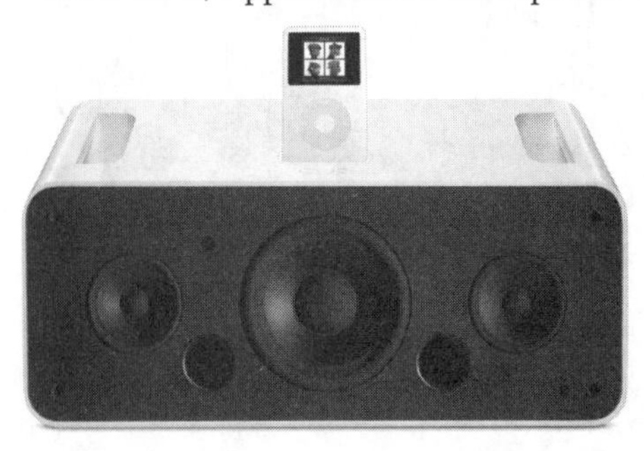

Powered speakers

Models include: **Audioengine 5; Genelec 8030A**
Cost (approx): **$350–1200**

One problem with the iPhone speaker units described opposite is that they offer no flexibility in terms of stereo separation – unlike traditional hi-fi speakers, which you can position as far apart as you like. Pairs of powered speakers – aka active speakers – get around this problem. One or both of the speakers contains its own amplifier, so all you need is a sound source – such as an iPhone or your computer.

The Audioengine 5 stands out as a very neat solution for both iPhones and iPods (pictured here with an iPod dock connected). These 70W-per-channel speakers feature a USB port up-top for charging and a separate minijack socket for the audio. For the price ($250), they sound excellent.

At the top end of the market are Genelec's range of bi-amplified speakers (that is, both speakers have an integrated amplifier). These have long been a popular choice for studio use, thanks to their incredibly detailed and rich sound, but they're equally well suited for audiophile home use. Each speaker connects using its own XLR cable, so you'll have to get a suitable two-channel preamp or DI box to create a "balanced" signal from your iPhone. The resulting sound is hugely impressive.

Hi-fi and speaker connectors

Models include: Belkin Stereo Link Cables
Cost (approx): $10–15

Whether you want to connect your stereo to your computer to digitize your vinyl collection (see p.143), or connect your iPhone to your hi-fi to play music in your living room, you'll probably need an RCA-to-minijack cable – also known as an RCA-to-3.5mm cable. These are very easy to find, though many won't fit into the iPhone's recessed headphone socket without an adapter.

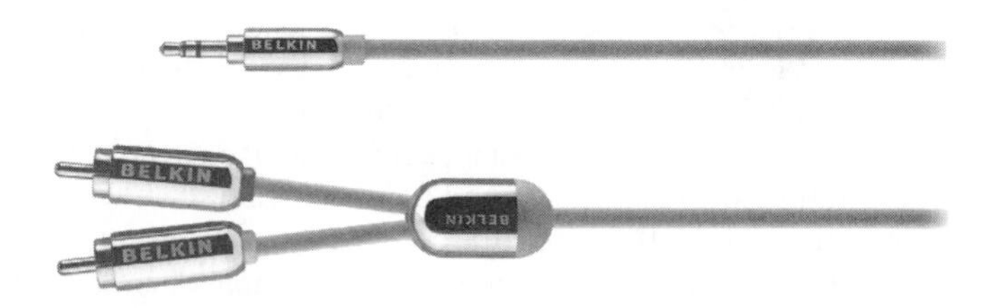

One example that will fit is the one made by Belkin, which also happens to come in an iPhone-friendly grey-silver colour scheme.

Belkin also produce a matching minijack-to-minijack cable, which can be useful for connecting your iPhone to line-in sockets on computers and speakers.

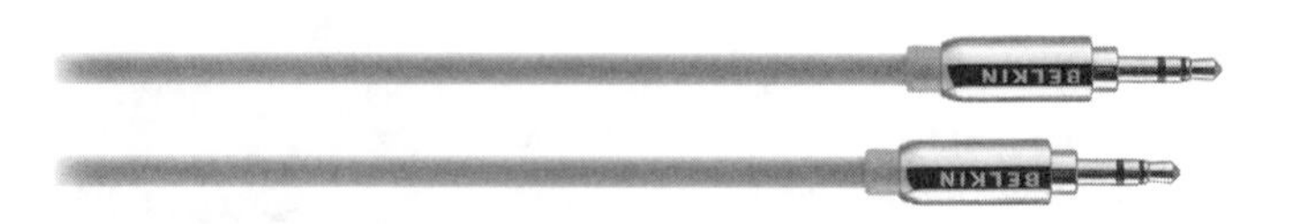

AirTunes

Models include: Apple AirPort Express
Cost (approx): $99

Many people connect an iPhone or iPod to their hi-fi for home listening. But it's often more convenient to play music direct from iTunes on your Mac or PC. Aside from anything else, there's room for a bigger music collection on your computer than on your iPhone.

If your hi-fi has a line-in socket, you can hook up your computer with an RCA-to-minijack cable (see opposite), but a neater solution is Apple's AirPort Express wireless base station, with its so-called AirTunes feature. Attach one of these to a power point near to your hi-fi and connect it to the stereo with an RCA-to-minijack cable. Then any computer with Wi-Fi can beam music straight from iTunes to the hi-fi. And what's more, with the right software (see p.164) you can use your iPhone as a Wi-Fi remote control for the music around your home.

AirPort Express can also act as a wireless router for Internet and printer sharing.

FM radio transmitters

Models include: Griffin iTrip
Cost (approx): $40 (not available in the UK)

These clever little devices turn your iPhone into an extremely short-range FM radio station. Once you've attached one, any radio within range (theoretically around thirty feet, though a few feet is more realistic to achieve a decent quality of sound) can then tune in to whatever music or podcast the iPhone is playing. The sound quality isn't as good as you'd get by attaching to a stereo via a cable (see previous page) and there can be interference, especially in built-up areas. But FM transmitters are very convenient and allow you to play your iPhone's music through any radio, including those – such as portables and car stereos – which don't offer a line-in socket.

Some models, such as the Griffin iTrip (pictured), which will soon be available for the iPhone, connect directly to the Dock socket and draw power from the phone's battery. Others contain their own battery and connect via the headphone jack – not as neat, but these types will work with any audio device.

FM transmitters were for a long time technically illegal in Europe, as they breached transmission laws. However, the law has changed, saving iPhone and iPod owners the hassle of importing them from the US.

Car accessories

Several major car manufacturers are now offering built-in iPod/ iPhone connectivity – among them BMW, Volvo, Mercedes and Nissan – which will allow you to control your music from your dashboard. Combine with a Bluetooth headset for safely receiving calls (see p.237) and you're all kitted out. But don't worry if you lack a recent high-end vehicle – you can also do things piecemeal:

Audio connectors

Unless your car stereo has a line-in socket, the two options are an FM transmitter (see opposite) or a cassette adapter. The latter will only work with a cassette player, of course, but they tend to provide better sound than an FM transmitter, and they're not expensive. There are various cassette adapters on the market, but the very cheapest ones have a tendency to produce wails and hisses. The Monster iCarPlay ($25) is one model that's both decent quality and compatible with the iPhone's recessed minijack socket.

Chargers

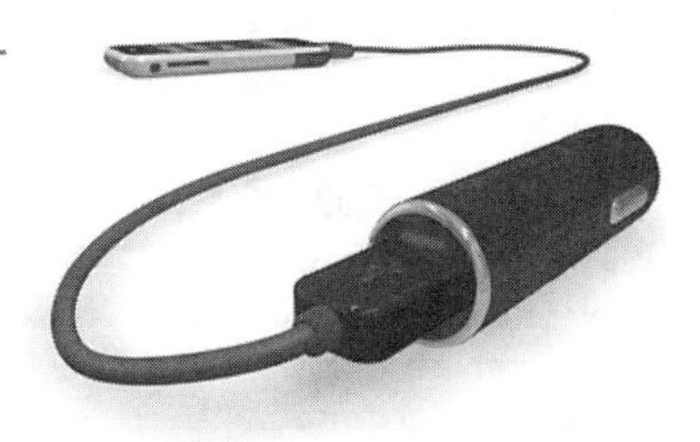

For in-car charging, try an XtremeMac's InCharge Auto Charger (pictured) or the Griffin PowerJolt Charger, both of which slot into a standard 12v car accessory socket.

For more auto iPhone solutions, visit iplaymycar.com

22

Websites

iPhones online

Within weeks of the iPhone's launch there were already scores of websites focusing on nothing else. Of course, all the existing iPod sites were also enjoying the new gadget. Following are some of the best sites for iPhone news, reviews, tips, troubleshooting advice, accessory reviews and forums where you can post queries.

News, reviews and how-tos

Everything iCafe everythingicafe.com
iLounge ilounge.com
iPhone Atlas iphoneatlas.com
iPhone Freak iphonefreak.com

Accessory stores

Amazon amazon.com or .co.uk
Apple Store apple.com/store
Griffin griffintechnology.com
iLounge ilounge.pricegrabber.com

Help

If you have an ailing iPhone or you want the latest tip, tap Safari and drop in to one of these support sites…

Apple Support apple.com/support/iphone
iPhone Atlas iphoneatlas.com
MacFixit macfixit.com

… or pose a question to one of the iPhone junkies who spend their waking hours on forums such as:

Apple Forums discussions.apple.com
Everything iPhone everythingicafe.com/forum
iLounge forums forums.ilounge.com
MacRumors forums.macrumors.com
Talk iPhone talkiphone.com

Hacks and mods

If you're the kind of person who likes to take things apart, check out the "Cracking Open the Apple iPhone" article at TechRepublic, and follow up with one of the numerous "iPhone disassembly" or "iPhone mod" posts on YouTube and Flickr. Trying any such thing at home will, of course, thoroughly void your warranty.

TechRepublic techrepublic.com
Flickr flickr.com
YouTube youtube.com

And to find out who's recently managed to get an iPhone to do what, drop into:

iPhone Hacks iphonehacks.com

Scripts and plug-ins

Apple Downloads apple.com/downloads
Doug's AppleScripts dougscripts.com/itunes
iLounge ilounge.com/downloads.php

Blogs

Add the following to your RSS page (see p.181) for an endless drop-feed of Apple titbits – including plenty on the iPhone.

Daring Fireball daringfireball.com
iPhone News iphonews.com
Secret Diary of Steve Jobs fakesteve.blogspot.com
Tao of Mac the.taoofmac.com
The iPhone Blog theiphoneblog.com

Feedback

If you have a suggestion about how the iPhone could be improved, then tell Apple at:

iPhone Feedback apple.com/feedback/iphone.html

23

iPhone weirdness

Staggering strangeness

It is, without a doubt, a weird world, and there's nothing quite like a new Apple gadget to bring the crackpots out of the cupboard. Even before anyone had seen an iPhone, the web was alive with crazy predictions; then there were the much-publicized queues in New York City in the days before the launch; and since then it has only got odder. What follows are a few dispatches from the dark world of iPhone obsession...

The predictions

Before the iPhone was launched, there was no shortage of ideas online for what it would look like, ranging from the nearly sane…

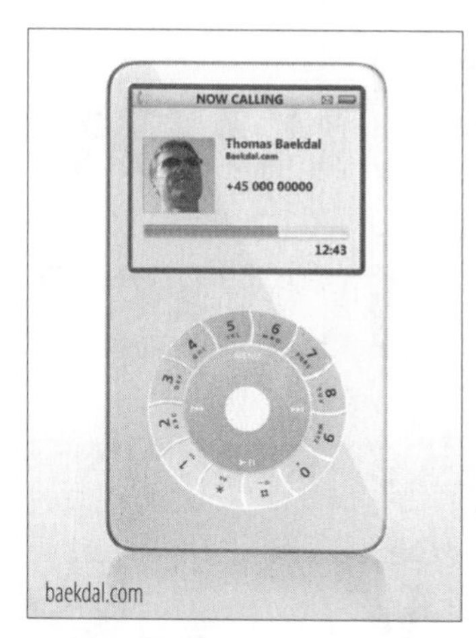

baekdal.com

Isamu Sanada

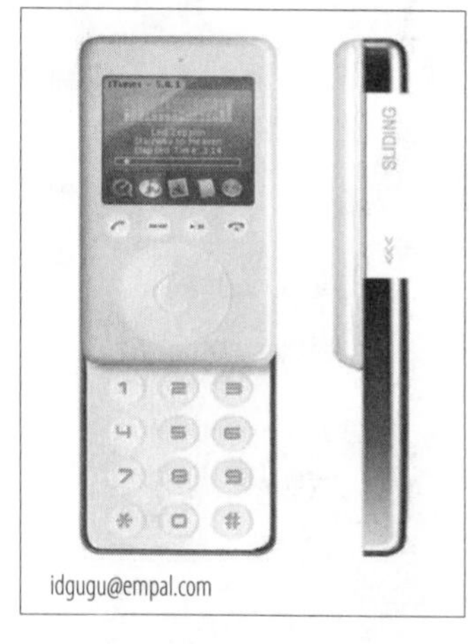

idgugu@empal.com

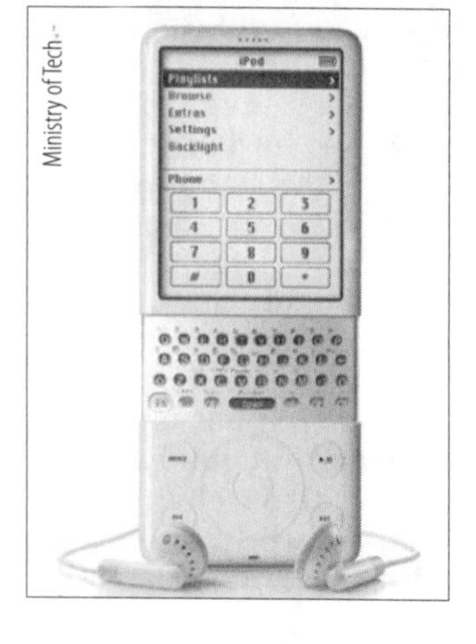

Ministry of Tech.

… to the completely insane. For more, see:

Product Dose productdose.com/article.php?article_id=5893
iPhone Concept Blog appleiphone.blogspot.com

There were even a few ideas that did more for the evolution of the rubber band and sticky tape than either iPods or phones…

The Apple phone that never was

Perhaps the strangest images to have circulated the web, however, are those illustrated below. Allegedly, they illustrate a phone for which Apple filed a patent back in 1982…

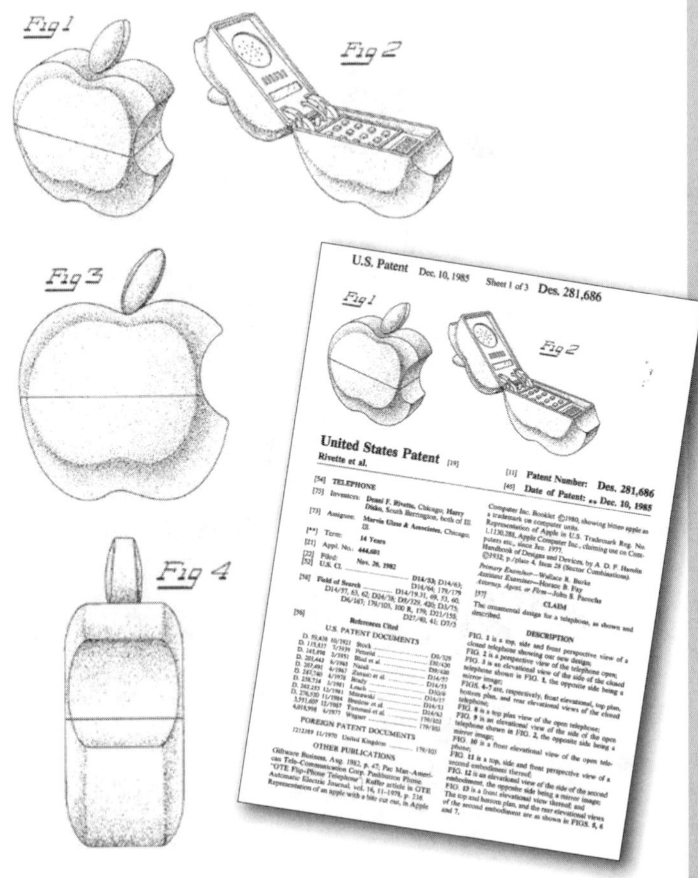

U.S. Patent Dec. 10, 1985 Sheet 1 of 3 Des. 281,686

United States Patent [19]
Rivette et al.

[11] **Patent Number: Des. 281,686**
[45] **Date of Patent: ** Dec. 10, 1985**

[54] TELEPHONE
[75] Inventors: Dean F. Rivette, Chicago; Harry Disko, South Barrington, both of Ill.
[73] Assignee: Marvin Glass & Associates, Chicago, Ill.
[**] Term: 14 Years
[21] Appl. No.: 444,601
[22] Filed: Nov. 26, 1982
[52] U.S. Cl. D14/53; D14/63; D14/64; 179/179
[58] Field of Search .... D14/19.31, 69, 53, 60, D14/57, 63, 62; D24/38; D8/329, 420; D3/75; D6/167; 179/103, 100 R, 179; D21/155, D27/40, 41; D7/5
[56] References Cited
U.S. PATENT DOCUMENTS
D. 59,438 10/1921 Stock
D. 115,837 7/1939 Petzold D9/320
D. 141,898 2/1951 Blad et al. D9/420
D. 201,443 6/1965 Natali D9/420
D. 207,491 4/1967 Zanuso et al. D14/57
D. 247,740 4/1978 Brady D14/53
D. 258,714 3/1981 Leach D10/6
D. 262,255 12/1981 Morawski D14/17
D. 276,520 11/1984 Breslow et al. D14/53
3,551,607 12/1967 Tomonari et al. D14/63
4,018,998 4/1977 Wagner 179/101
.... 179/103
FOREIGN PATENT DOCUMENTS
1212389 11/1970 United Kingdom 179/103
OTHER PUBLICATIONS
Giftware Business, Aug. 1982, p. 47, Pac Man–American Tele-Communication Corp. Pushbutton Phone. "GTE Flip-Phone Telephone"; Ruffer article in GTE Automatic Electric Journal, vol. 16, 11-1978, p. 216. Representation of an apple with a bite cut out, in Apple Computer Inc. Booklet ©1980, showing bitten apple as a trademark on computer units.
Representation of Apple in U.S. Trademark Reg. No. 1,1130,288, Apple Computer Inc., claiming use on Computers etc., since Jan. 1977.
Handbook of Designs and Devices, by A. D. F. Hamlin ©1932, p./plate 4, Item 28 (Sector Combinations).

Primary Examiner—Wallace R. Burke
Assistant Examiner—Horace B. Fay
Attorney, Agent, or Firm—John S. Pacocha

[57] CLAIM

The ornamental design for a telephone, as shown and described.

DESCRIPTION

FIG. 1 is a top, side and front perspective view of a closed telephone showing our new design;
FIG. 2 is a perspective view of the telephone open;
FIG. 3 is an elevational view of the side of the closed telephone shown in FIG. 1, the opposite side being a mirror image;
FIGS. 4-7 are, respectively, front elevational, top plan, bottom plan, and rear elevational views of the closed telephone;
FIG. 8 is a top plan view of the open telephone;
FIG. 9 is an elevational view of the side of the open telephone shown in FIG. 2, the opposite side being a mirror image;
FIG. 10 is a front elevational view of the open telephone;
FIG. 11 is a top, side and front perspective view of a second embodiment thereof;
FIG. 12 is an elevational view of the side of the second embodiment, the opposite side being a mirror image;
FIG. 13 is a front elevational view thereof; and
The top and bottom plan, and the rear elevational views of the second embodiment are as shown in FIGS. 5, 6 and 7.

Wool iPhone

The iPhone for minors and knitwear fanatics:
daddytypes.com

Cake iPhone

The mobile you can eat between meals without ruining your appetite:
mobilewhack.com/iphone-cake---cool

Lightning warning!

If your iPhone obsession is so severe that you wear your earbuds or headset whatever the weather, you might be risking your life. Indeed, one thing your weather widget won't tell you is that wearing earphones during a thunderstorm could increase the damage caused if you're unlucky enough to be struck by lightning. The metal can channel the energy into your ear, causing, in one instance, "… wishbone-shaped chest and neck burns, ruptured eardrums and a broken jaw". You have been warned.

Cotton and silk

If your desire for an iPhone isn't matched by your bank balance, tide yourself over with a paper version, courtesy of: homepage.mac.com/colin-baxter/ipod/ipodclick.html

If money *isn't* an issue, on the other hand… gizmodo.com/gadgets/igems

iPhone shuffle

There are several incarnations of this theme doing the rounds … a screenless phone whose USP is its ability to call and text your friends and family at random. blog.thesedays.com

iPhone generator

Worried about your iPhone's battery lifespan? You could take the advice on p.88, or just plonk for an iPhone generator for power emergencies. Watch this space for the hotly anticipated hydro-electric version (dam not supplied): andrewsavory.com

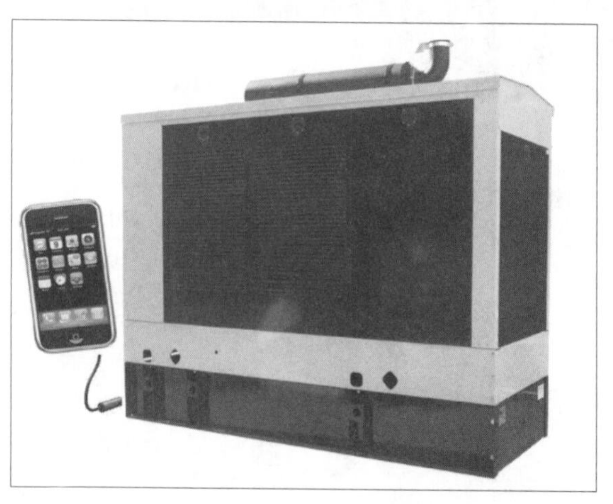

iPhone size

An essential reference tool if you ever need to communicate the size of your iPhone to a stranger but do not happen to have said digital gizmo to hand: iphonesize.com

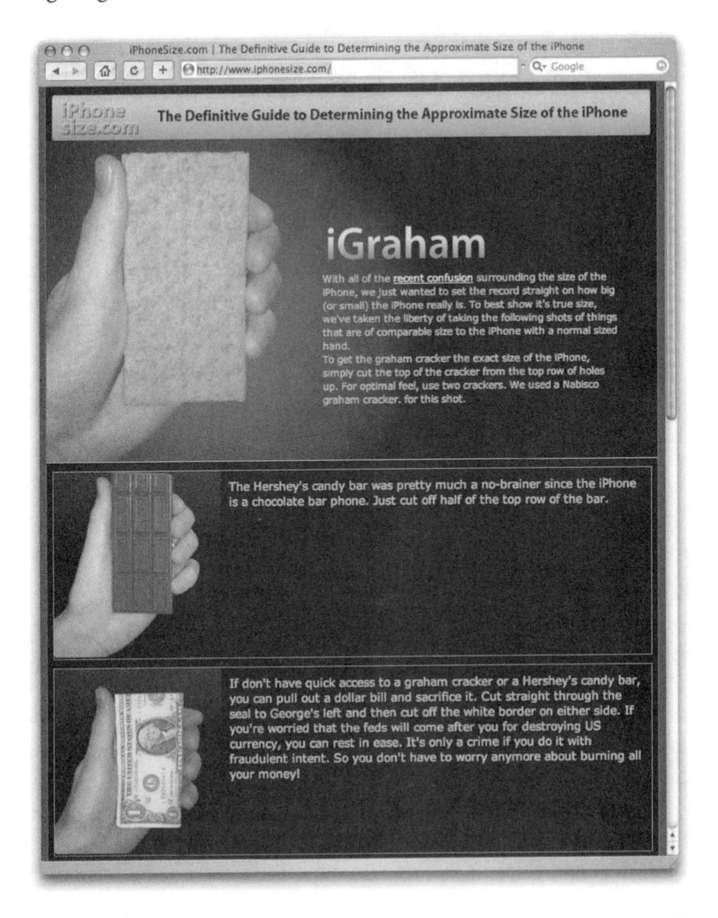

Engraving

Apple have not yet offered an engraving service for the iPhone, but the guys at DeviceNineSix.com have...

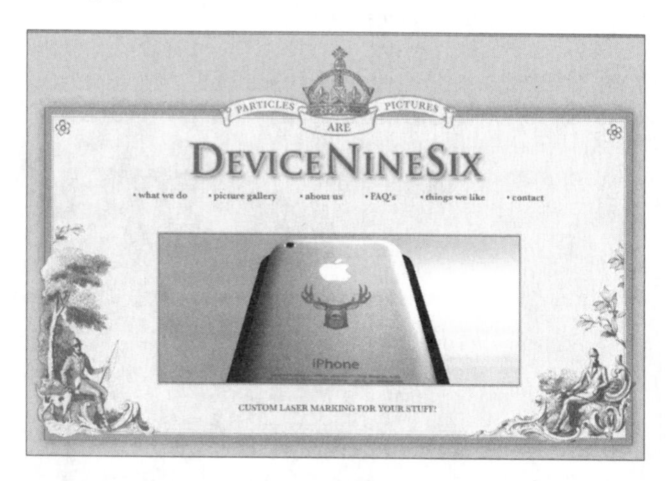

Lego iPhones

Compared to the wondrous Lego creations that can be found online paying tribute to the humble iPod, those for the iPhone are rather second-rate. Admittedly it is hard to create a flat-screened device using a medium that is so fundamentally knobbly, but surely someone can do better than this?

flickr.com/photos/gilest

YouTube

Thankfully, there's more iPhone (and Lego) nonsense on YouTube to keep us all entertained. The "2001: A Space iPhone" and "Conan iPhone Commercial" clips are both essential viewing…

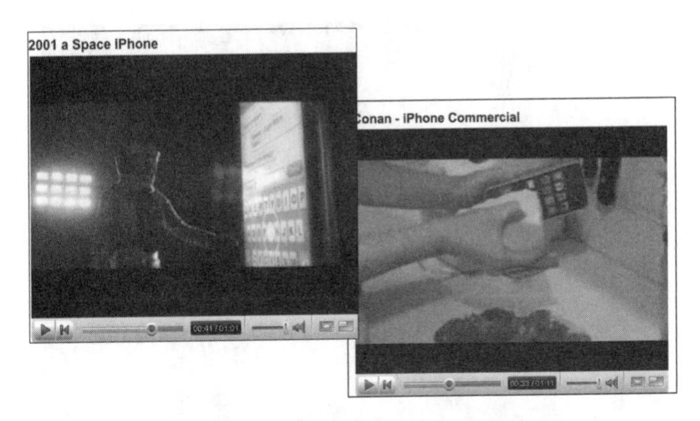

…and these animations are guaranteed to raise a smile:

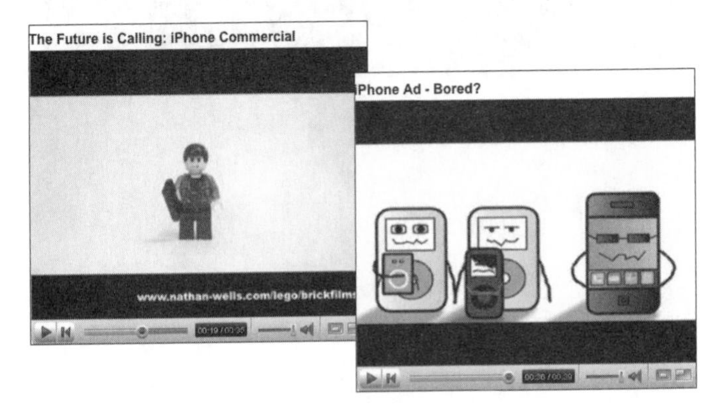

Will It Blend?

Who on Earth would want to use an iPhone as an iPhone when there's the option of turning it into a smoothie instead?

willitblend.com

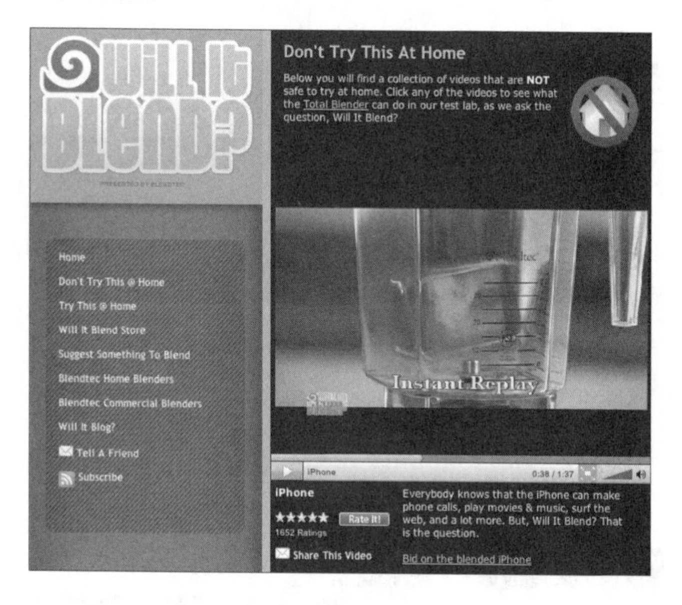

Index

A

AAC 138, 139, 155
Access Lists 67
accessories 28, 235
AccuRadio 165
activation 48
adapters 96
Address Book 97, 98, 100, 129
Adobe Digital Editions 219
aggregators 17
AgingBooth 205
AIFF 138, 139
AIM 129, 133
Air Display 231
Airplane mode 64
AirPort Express 243
Air Sharing HD 231
AirTunes 243
alerts 52
Amazon 246
Android 13
answering calls 115
Antenna-gate 6
AnyDVD 140
AOL 27, 103, 152, 187, 191
AppleCare 90
Apple Forums 247
Apple headset 113
Apple ID 48, 152
Apple logo 84
Apple Lossless 89, 138, 139
Apple Refurb Store 35
AppleScripts 248
Apple Store 34, 246
Apple Support 247
apps 13, 71
    deleting 82
    downloading 72
    folders 81
    multitasking 77
    organizing 80
    rating 82
    syncing 49, 74
    updating 75
At Bat 2010 171

AT&T 20, 34
Audacity 144
Audible 138
audiobooks 32, 163
audio editing software 143
audio settings 57
authorized computers 155
authorizing computers for iTunes 73
auto-capitalization 61
auto-correct 60
AutoFill 179
Auto-Lock 55
AV cables 169
Awesome Note 227

B

backed up settings 84
backing up 86
batteries 18, 47, 65, 87, 256
battery life, maximizing 88
battery replacement 89, 90
BBC iPlayer 171
billing 48
Bing 176
birthdays 107
bitrates 138
blenders 260
Bloom 232
Bluetooth 19, 45, 64, 68, 88, 96
Bluetooth headsets 237, 245
Boingo 66
bookmarking in iBooks 218
bookmarks 26, 52
  Safari 178
books, syncing 50
Boxee 172
Braun 230
browsers 185
buying an iPhone 31, 33

C

cached, Google 181
cache, Safari 176
cake 254
calculator 230
Calendar 224
calendars 49, 97
Calengoo 225
caller ID 123
call forwarding 123
CallHer & CallHim 124
callLog 124
Call Options 113
calls 109, 110, 131
  to an email address 126
  via the Internet 125
camera 201
  removal of 203
  taking better pictures 202
Camera+ 205
CameraBag 205
Camera Roll 204
Caps Lock 59
car accessories 245
carphones 64
cases 238
cassettes, importing into iTunes 143
CDs 29
charge cycles 87
chargers 35, 245

charging 47
Chat 133
Clock 226
comic books 221
computers 16
  requirements 16
  upgrading 16
conference calling 111
connecting 64
contacts 49, 95, 111
  adding pictures 101, 108
  editing on a computer 100
  editing on the iPhone 107
  importing from an old phone 96
  photographing 204
  syncing with the iPhone 102
contracts 20
converting file formats 139
converting video files 142
Cookies 180
copying music, legality 30
Copy & Paste 61
copy protection, DVDs 140
copyright 30
corporate Exchange email 189
Cover Flow 13, 159
crashes 83
custom ringtones 53
Cut & Paste 61

D

Daring Fireball 248
DataPilot 99
declining calls 114
deleting apps 82
deleting music 162
desktops 17
diagnostic codes 91
dialling codes 126
Dial Plate 124
dictionary 60
Dictionary in iBooks 218
digital rights management *See* DRM
directions (maps) 212
disconnecting 48
DMP3 29
Dock 15, 46
Dock connector socket 45
Doc² 228
.Mac 27
Doug's AppleScripts 248
downloading music 30
DRM 30, 155
Dropbox 231
Dumas, Daniel 20
DVDs 30, 86

E

earphones 29
eBay 79
EDGE 25, 64, 68
eject button 48
Elgato 172
email 27, 52, 187
  default account 196
  jargon buster 188
  managing sent mail 195, 198
  message preview 195
  missing messages 197
  POP 188

problems 197
sending and receiving 197
settings 195
setting up 187
signatures 196
emergency calls 110
Emoji 132
emoticons 132
engraving 258
Enhanced Data rates for GSM Evolution *See* EDGE
Eno, Brian 232
Entourage 102
calendars 224
EQ 89, 160
Etude 166
Evernote 227
Excel documents 14, 184
Exchange Accounts 49, 69, 188, 225
Exchange servers 69

F

Facebook 79, 186
FaceDialer 124
FaceTime 111, 120, 201
FairPlay DRM 155
Fast app switching 76
Fast DVD Copy 140
FatBooth 205
Favorites 106, 110
Favorites list 86
fetch services 190
file sharing 30
Financial Times, The 223
firmware 32, 67
update 85
Flash 26
flash photography 203
Flickr 79, 208, 247
FM radio transmitters 244
folders 81
forgetting networks 65
forums 247
free calls 125
Freecom 172
Fring 127

G

GarageBand 53, 144
GeeTasks 226
Genius 155, 161
German functionalism 230
gift cards 154
gigabytes 32
Global Positioning System *See* GPS
gloves 19
Gmail 27, 79, 103, 187, 191
GoodReader 219
Google 176, 218, 225
Android 13
Google Accounts 49
Google Docs 228
Google Local 211
Google Mail *See* Gmail
Google Maps 209
Google Street View 211
Google Talk 133
Googling tips 182
Gourmet Egg Timer 232
GPRS 25

GPS 77, 210
graphic novels 221
graphs 230
GSM 64, 68
Guardian, The 223
GuitarToolKit 166

H

H.264 142
hacks 247
HandBrake 140
Harbor Compass Pro 214
hard drives 18
Hauppauge 172
headphone adapters 236
headphones *See* earphones
headsets, Bluetooth 237
headsets, wired 236
hi-fi
  connecting to 166, 242
  recording from 143
highlighting in iBooks 218
Hipstamatic 205
Historic Earth 214
history, Safari 176
hold, putting calls on 111, 115
Home button 45, 113, 161
Home Screen 54, 80
Hotmail 103
hotspots 64, 66

I

iBooks 215
iBookstore 215
iCal 224
iCarPlay 245
iChat 133
icons (webclips) 78
ICQ 133
iDisk 231
IE *See* Internet Explorer
IEEE 802.11x *See* Wi-Fi
ignoring calls 115
images 201
iMangaX 221
IMAP 188, 189
iMovie 204
importing CDs 137
importing DVDs 140
Info (syncing) 49
Instapaper 222
insurance 90
Internet access 24, 64
Internet audio 41
Internet Message Access Protocol 188
Internet tools 173
iOS Maps 214
iPad 11, 73
iPhone Shuffle 256
iPhonesize.com 257
iPhoto 206, 208
iPlayer 171
iPod 11, 13, 77
  accessories 28
  batteries 18
  iPhone as 135
  nano 18
  shuffle 18
  touch 18

iPod app 156
iPod controls 157
iRingtoner 53
iRip 147
iSync 97
iToner 53
iTopoMaps 214
iTrip 244
iTunes 15, 40
  Internet audio content 41
  multiple windows 43
  preferences 41, 43
  tips 43
  what is it? 15
iTunes Store 15, 16, 28, 53, 149
iTunes Store Account 73, 152, 216
iTunes U 154
iVerse Comics 221

J

jailbreaking 21
Jajah 127
Jangl 126
Java 26
JavaScript 26 180
Jobs, Steve 248
Jotnot Scanner Pro 231

K

keyboard 14, 130
Kindle 221
knitwear 254

L

L2TP 70
laptops 17, 27, 129
latency 67
Lego 258
Location Services 210
Lock Screen 54
LogMeIn 70
Looptastic Producer 166
lyrics 160

M

MAC addresses 67
Mac computers 11
MacFixit 247
Mac Rumors 33
Mail 131
Mail app 192
manga 221
Manually Manage Music and Video 148
Maps 65, 113, 131, 209, 227
  traffic conditions 213
Marvel 221
megabytes 32
memory cards 18
merging calls 112
messaging 133
Methodology – Creative Inspiration Flashcards 232
Met Office 229
mic 45
microscope, iPhone as 202
MicroSIM cards 21
Microsoft Exchange 187, 188, 196

See Exchange Accounts
Microsoft Exchange servers 69
minidiscs, importing into iTunes 143
minijack socket 44
missed calls 115
miTypewriter 63
MMS 108
Mobile Master 99
MobileMe 27, 48, 49, 152, 187, 191, 208, 225, 231
mono headsets 69
Motion JPEG 142
movie trailers 154
MP3 41, 138, 139
MPEG-4 142
MSN 133
multiple calls 112
multitasking 77
music 51
 apps 166
 controls 157
 copying from computer to computer 147
 copying from iPhone to computer 147
 formats 89
 managing 40, 162
 settings 160
 sharing 147
 syncing 50
Music Rescue 147
mute 111
MySpace 79

N

naming your iPhone 54
Napster 103
National Geographic World Atlas HD 214
NetNewsWire 222
NetPortal 70
news apps 222
newsfeeds 222
NewsRack 222
NLog Free Synth 166
Notepad (PC) 98
Notes 58, 227
 alternatives 227

O

O2 20, 34
Oblique Strategies 232
Office² 230
Opera Mini 185
operating systems 14, 32
optimized sites 79, 184
OS X 14, 16
Outlook 98, 100, 129
 calendars 224
Outlook Express 100
overseas, using iPhone 22, 117
Oxfam Recycle Scheme 36

P

Pandora 165
Pano 205
Parental controls 56

Parker, Sean 103
Passcode Lock 55, 152, 179
passwords in Safari 179
Paste 61
Pastebot 62
patents 253
pay-as-you-go 22
PCalc Lite 230
PDFs 184, 218
photos 50, 201
  contacts 204
  importing from Mac or PC 206
  sideshows 208
  syncing 50
  viewing 207
Photoshop 205
Plane Finder 214
Plaxo 103
play controls 77
playlists 50, 145, 161
Pocket CV 228
podcasts 17, 40, 50, 51, 151, 163, 244
  on the iPhone 151
  subscribing 151
PodServe 29
Pogo Stylus 19
Point Inside 214
POP 188
POP3 188
pop-up blocker 180
Post Office Protocol 188
PowerPoint 14
PPTP 70
pre-pay 22
private browsing 185
projector, connecting to 169
PS Mobile 205
punctuation 58
push services 190

Q

quad-band 22
Quick Graph 230
QuickTime Pro 172
Quicky Browser 185

R

Rams, Dieter 230
rating apps 82
RCA cables 242
reading apps 215
rebooting 84
Rebtel 128
recent calls 105
Recents 110, 115
reception issues 6
recycling 36
redialling 110
refurbished iPhones 35
refurbished iPods 35
Remember The Milk 226
remote access 70
Remote (app) 164
remote control 164
renting movies 153
repairs 91
replace text 63
Reset Network Settings 84
resetting 84
restoring 84

Restrictions 56
returning calls 118
ringtones 52, 155
    custom 53
RipDigital 29
ripping CDs 29
roaming 22, 23, 117
routers 67
Rowmote Pro 162, 164
RSS 181, 222

S

Safari 26, 61, 113, 131, 175
    basic controls 175
    multiple pages 177
    problems 181
    searching for text 181
    security 179
    settings 180
satellite maps 212
Satellite view 212
Schmidt, Peter 232
screen 19
screen brightness 88
screen protectors 238
Screen zoom 56
scrubbing 158
searching 57
secondhand iPhones 35
security (Safari) 179
sending email 192
Service Pack 2 16
service plans 48
settings
    accessibility 56
    apps 74
    Bluetooth 69
    brightness 88
    call forwarding 123
    email 187, 195
    email accounts 191
    EQ 89
    Erase All Content and Settings 84
    general 55
    iBooks 217
    importing music 138
    iTunes Store 152
    keyboard 61
    movies 169
    music 160
    push 190
    reset 84
    Reset Keyboard Dictionary 60
    Reset Network Settings 84
    restrictions 153
    sounds 52, 113
    video 169
    voicemail 117
    VPN 70
    Wi-Fi 65
setting up 39, 52
Shake to shuffle 160
Shake to undo 63
Share contact 108
Shazam 165
Sheet² 230
sheet music 166
Ship Finder 214
Short Message Service *See* SMS
Sideshows 208
silencing a call 114
silent 53

Silent ringer switch 44
SIM cards 21, 22, 110
Simple Passcode 55
SIM readers 97, 98
SIM tray 45
Single-Molecule Spectroscopy 130
Skype 125, 129
sleep, iPod app function 159
Sleep/Wake button 45, 84, 88, 114
smartphones 11
Smart Playlists 145
SMS 55, 118, 130
  from a computer 134
  internationally 134
SMS Preview 55
social security numbers 22
software problems 83
Software Update 16, 40
Sony Reader 220
Sound Check 160
SoundHound 165
sound quality 28, 89
Source list 42
speaker 44, 111
speaker-phone 111, 119
speakers
  connectors 242
  powered speakers 241
  speaker units 240
speed tests 67
speed-typing 59
spell check 60
spoken word 163
Spotify 165
Spotlight Search 57
Stanza 221
star ratings 161
Status bar 45
Stocks 229
stopwatch 226
storage capacity 32
stylus 19
Super Caller ID 123
super cookies 79, 180
switching apps 76
symbols, typing 58
syncing 26, 49
  Account Information 52
  apps 49
  audio and video files 146
  books 50
  calendars 189
  contacts 189
  forced 51
  iBooks 219
  multiple computers 51
  with Microsoft Exchange 189
  with MobileMe 189
SyncYourMail 189
System Profiler 16

T

tabbed browsing 185
Tao of Mac 248
TeamViewer 70
TechRepublic 247
Terrain view 212
tethering 27
threads (in emails) 196
Tiltshift 205
timer 226

T-Mobile 66
Text Edit (Mac) 98
texting 130
  from a computer 134
  internationally 134
3G 25, 64, 68
Thunderbird 103
todo lists 226
Toodledo 226
touch screen 19
Transit Maps 214
troubleshooting 83
Truphone 126
TuneIn Radio 165
TV
  connecting to 169
  recording from 172
  watching 171
TVCatchup 171
Twitter 222
typing 58
typing tests 61
typos 14

U

Universal Dock 46
unlocking 21
Unofficial Apple Weblog (TUAW) 91
USA Today 223
USB phones 129
USB/USB2 15, 17
USB cables 35, 46

V

vCards 97, 98, 108
VGA Adapter 169
vibrate 53, 113
video 50, 51, 167
  calls 120
  editing 204
  formats 142
  recording from TV 172
  settings 57
  shooting 203
Video Camera (for iPhone 2G and 3G) 205
video out 169
vinyl, importing into iTunes 143
Vista 16
Visual Voicemail 116
VLC Remote 164
Voice Control 122
voicemail 116
voicemail password 116
Voice Memos 227
VoiceOver 56
voice plans 20
VoIP 77
volume buttons 44, 114
Volume Limiter 160
VPN 69

W

Walkmans, recording from 143
wallpaper 54
Wall Street Journal, The 223
warranties 90
WAV 138

Wayport 66
Weather 229
weather apps 229
web apps 79, 185
web app directory 79
web browsers 26
web browsing 14
webclips 71, 78
websites
  directories 79
  help 247
  mobile versions of 79
  news 246
weirdness 249
WEP Hex 66
White on Black 57
Wi-Fi 45, 64, 65, 88, 113, 243
WiFinder 66
Wikipedia 218
Windows 14, 16
Windows Live Messenger 129
Windows Update 16
WinFonie Mobile 2 99
wired headsets 236
Word documents 14, 184
WordPress 228
WPA 67
WriteRoom 228

X

Xmarks 179
XP 16, 98

Y

Yahoo! 27, 176, 187, 190, 191
Yahoo! Address Book 99, 107
Yahoo! Mail 99
Yahoo! Messenger 129
YouTube 170, 247, 259

Z

Zinio 223
ZONE Finder 66
zoom photography 203